Electricity from Sunlight

Electricity from Sunlight

An Introduction to Photovoltaics

Paul A. Lynn BSc(Eng), PhD

formerly
Imperial College London

WILEY

A John Wiley & Sons, Ltd., Publication

This edition first published 2010
© 2010, John Wiley & Sons, Ltd

Registered office
John Wiley & Sons Ltd, The Atrium, Southern Gate, Chichester, West Sussex,
PO19 8SQ, United Kingdom

For details of our global editorial offices, for customer services and for information
about how to apply for permission to reuse the copyright material in this book
please see our website at www.wiley.com.

Library of Congress Cataloging-in-Publication Data
Lynn, Paul A.
    Electricity from sunlight : an introduction to photovoltaics / Paul A. Lynn.
        p. cm.
    Includes bibliographical references and index.
    ISBN 978-0-470-74560-1 (cloth)
    1. Photovoltaic power generation. 2. Solar cells. 3. Solar energy. I. Title.
    TK1087.L96 2010
    621.31′244–dc22
                                                                    2009054395

A catalogue record for this book is available from the British Library.

ISBN: 978-0-470-74560-1

Set in 10/12 pt Times New Roman by Toppan Best-set Premedia Limited
Printed in Singapore by Fabulous Printers Pte Ltd

Contents

About the author

Paul A. Lynn obtained his BSc(Eng) and PhD degrees in electrical engineering from Imperial College London. After several years in the electronics industry he lectured at Imperial College and the University of Bristol. In 1993 he was appointed founding Managing Editor of the Wiley international journal *Progress in Photovoltaics: Research and Applications*, which he managed for 14 years. His previous publications include 5 text-books, more than 50 technical papers and articles, and 2 short books on English country pubs. In his spare time Paul maintains a strong interest in environmental matters and is a long-time member of *Friends of the Earth*. He has designed and built three prize-winning solar catamarans and in 2003 he and his wife Ulrike made the first-ever solar-powered voyage along the entire River Thames from Gloucestershire to London.

Preface

Photovoltaics (PV), the 'carbon-free' technology that converts sunlight directly into electricity, has grown dramatically in recent years. Unique among the renewable energies in its interaction with the built environment, PV is becoming part of the daily experience of citizens in developed countries as millions of PV modules are installed on rooftops and building facades. People living in sunshine countries will increasingly live in solar homes or receive their electricity from large PV power plants. Many governments around the world are now keen to promote renewable electricity as an essential part of the 21st century's energy mix, and PV is set for an exciting future.

This book is designed for students and professionals looking for a concise, authoritative, and up-to-date introduction to PV and its practical applications. I hope that it will also appeal to the large, and growing, number of thoughtful people who are fascinated by the idea of using solar cells to generate electricity and wish to understand their scientific principles. The book covers some challenging concepts in physics and electronics, but the tone is deliberately lighter than that of most academic texts, and there is comparatively little mathematics. I have included many colour photographs, gathered from around the World, to illustrate PV's huge and diverse range of practical applications.

In more detail, Chapter 1 introduces PV's scientific and historical context, suggests something of the magic of this new technology, and summarises its current status. The treatment of silicon solar cells in Chapter 2 includes material in semiconductor physics and quantum theory, described by a few key equations and supported by plenty of discussion. The new types of thin-film cell that have entered the global PV market in recent years are also introduced. Chapter 3 covers the characteristics of PV modules and arrays, discusses potential problems of interconnection and shading, and outlines the various types of system that track the sun, with or without concentration. The two major categories of PV system, grid-connected and stand-alone, provide the material for Chapters 4 and 5 respectively, and Chapter 6 concludes the story with some of the most important economic and environmental issues surrounding PV's remarkable progress.

Photovoltaic technology seeks to work with nature rather than to dominate or conquer it, satisfying our growing desire to live in tune with Planet Earth. I trust that this book will inspire as well as inform, making its own small contribution to an energy future increasingly based on 'electricity from sunlight'.

Paul A. Lynn

Butcombe, Bristol, England

Spring 2010

Acknowledgements

There is nothing like a good set of pictures to illustrate PV's extraordinary progress and I have enjoyed enlivening the text with colour photographs obtained from around the world. I hope that my readers will regard them as an important and inspirational aspect of the book. They come from widespread sources and I have received generous cooperation from people in many organisations and companies who have provided copyright permissions and, in several cases, suggested stunning alternatives to illustrate particular topics.

I am especially grateful to the two international organisations that have provided the lion's share of the photographs reproduced in this book:

1. The European Photovoltaic Industry Association (EPIA)

The EPIA is the world's largest industry association devoted to Photovoltaics, with more than 200 business members representing about 95% of the European PV industry. EPIA members are active across the whole field of PV, from silicon producers, cell and module manufacturers, to system providers. Amongst the Association's many activities promoting a higher awareness and penetration of the technology, it represents the European PV industry in contact with political institutions and key decision makers.

The Association's informative website (www.epia.org) includes an excellent photo gallery with a comprehensive selection of images provided by business members. The author acknowledges use of the following photographs, which are reproduced by permission of the EPIA:

Figures 1.3, 1.12, 2.1, 2.3, 2.22, 2.23, 2.29, 3.1, 3.16, 3.18, 4.9, 4.11, 4.12, 4.14, 4.15, 4.16, 4.17, 4.18, 4.20, 4.21, 4.23, 4.24, 4.25, 4.26, 4.27, 5.2, 5.25, 5.27, 5.28, 5.33, 5.34, 5.35, 5.36, 5.37, 6.3, 6.4, 6.5, 6.6, 6.7, 6.8, 6.9, 6.10, 6.11, 6.12, 6.14, 6.16

Permission to use two of these photographs (Figures 1.3 and 4.20) for the front cover is also gratefully acknowledged.

At a personal level it is a pleasure to thank Michel Bataille, IT Manager of the EPIA, for his advice and technical assistance.

2. The International Energy Agency Photovoltaic Power Systems Programme (IEA PVPS)

The International Energy Agency (IEA), founded in 1974 as part of the Organisation for Economic Co-operation and Development (OECD), encourages energy cooperation among member countries. Its Photovoltaic Power Systems Programme (IEA PVPS), begun in 1993, now has 23 members worldwide and organises international projects to accelerate the development and deployment of Photovoltaics.

The IEA PVPS website (www.iea-pvps.org) includes a series of excellent Annual Reports giving up-to-date information about PV developments in the various member countries. The author acknowledges use of the following photographs from these Annual Reports, which are reproduced by permission of IEA PVPS.

Figures 1.1, 1.4, 1.8, 1.13, 1.14, 1.15, 2.2, 2.27, 3.3, 3.14, 4.2, 4.3, 4.6, 4.7, 4.10, 4.13, 4.19, 4.22, 4.28, 4.30, 4.31

Permission to use three of these photographs (Figures 1.4, 1.13, and 4.6) for the front cover is also gratefully acknowledged.

At a personal level it is a pleasure to thank Mary Brunisholz, Executive Secretary of IEA PVPS, for her enthusiastic advice and encouragement.

3. Additional acknowledgements

I am also grateful to a further group of companies and organisations that have agreed to their photographs appearing in this book, and for help received in each case from the named individual:

Amonix Inc. (Nate Morefield) 3425 Fujita Street, Torrance, CA 90505, USA	(Figure 3.21)
Boeing Images (Mary E. Kane), USA www.boeingimages.com	(Figure 2.32)
Dyesol Ltd (Viv Tulloch) P.O. Box 6212, Queanbeyan, NSW 2620, Australia	(Figure 2.35)
Dylan Cross Photographer (Dylan Cross), USA dylan@dylancross.com	(Figure 5.31)
First Solar Inc. (Brandon Michener) Rue de la Science 41, 1040 Brussels, Belgium	(Figure 2.31)
Isle of Eigg Heritage Trust (Maggie Fyffe) Isle of Eigg, Inverness-shire PH42 4RL, Scotland	(Figure 5.23b)
Padcon GmbH (Peter Perzl) Prinz-Ludwig-Strasse 5, 97264 Helmstadt, Germany	(Figure 4.5)
Steca Elektronik GmbH (Michael Voigtsberger) Mammostrasse 1, 87700 Memmingen, Germany	(Figures 5.9 and 5.12)
Tamarack Lake Electric Boat Company (Montgomery Gisborne) 207 Bayshore Drive, Brechin, Ontario L0K 1B0, Canada	(Figure 5.30)
Wind and Sun Ltd (Steve Wade) Leominster, Herefordshire HR6 0NR, England	(Figures 5.23(a) and 5.24)

The publishers acknowledge use of the above photographs, which are reproduced by permission of the copyright holders.

The use of three photographs (Figures 5.18, 5.19 and 5.20) from the NASA website, and several pictures from the Wikipedia website (Figures 1.5, 1.9, 1.10 and 5.32), is also gratefully acknowledged.

The author of a comparatively short but wide-ranging book on PV – or any other technology – inevitably draws on many sources for information and inspiration. In my case several longer and more specialised books, valued companions in recent years, have strongly influenced my understanding of PV and I freely acknowledge the debt I owe their authors, often for clear explanations of difficult concepts that I have attempted to summarise. These books are included in the chapter reference lists, and you may notice that a few of them appear rather frequently (especially items 1, 2, 3, and 4 in the reference list to Chapter 5). I have tried to give adequate and appropriate citations in the text.

My previous books on electrical and electronic subjects have been more in the nature of standard textbooks, illustrated with line drawings and a few black-and-white photographs. When the publishers agreed to my proposal for an introductory book on PV containing full-colour technical drawings and photographs, I realised that a whole new horizon was in prospect, and have enjoyed the challenge of trying to choose and use colour effectively. The photographs, many of them superb, have already been mentioned. It has also been a great pleasure to work closely with David Thompson, whose ability to transform my sometimes rough sketches into clear and attractive technical drawings has been something of an eye-opener. Dave's design sense and attention to detail are reflected in the many excellent colour figures that (as I trust my readers will agree) adorn the pages of this book.

For nearly 15 years my main involvement with PV was as Managing Editor of the Wiley international journal *Progress in Photovoltaics: Research & Applications*. Among the many editorial board members who gave valuable advice over that period, I should particularly like to mention Professor Martin Green of the University of New South Wales (UNSW), world-renowned for his research and development of silicon solar cells; and Professor Eduardo Lorenzo of the Polytechnic University of Madrid (UPM), whose encyclopaedic knowledge of PV systems and rural electrification was offered unstintingly. It was both a privilege and a pleasure to work with them for many years. And although any shortcomings in this book are certainly my own, any merits are at least partly due to them and other members of the board.

Finally I should like to thank the editorial team at Wiley UK in Chichester, and especially Simone Taylor, Nicky Skinner, Laura Bell, and Beth Dufour, for their enthusiasm and guidance during this project. They, and others, have eased into publication this account of an exciting new technology that magically, and quite literally, produces electricity from sunlight.

Paul A. Lynn

1 Introduction

1.1 The sun, earth, and renewable energy

We are entering a new solar age. For the last few hundred years humans have been using up fossil fuels that took around 400 million years to form and store underground. We must now put huge effort – technological and political – into energy systems that use the Sun's energy more directly. It is one of the most inspiring challenges facing today's engineers and scientists and a worthwhile career path for the next generation. Photovoltaics (PV), the subject of this book, is one of the exciting new technologies that is already helping us towards a solar future.

Most politicians and policymakers agree that a massive redirection of energy policy is essential if Planet Earth is to survive the 21st century in reasonable shape. This is not simply a matter of fuel reserves. It has become clear that, even if those reserves were unlimited, we could not continue to burn them with impunity. The resulting carbon dioxide emissions and increased global warming would almost certainly lead to a major environmental crisis. So the danger is now seen as a double-edged sword: on the one side, fossil fuel depletion; on the other, the increasing inability of the natural world to absorb emissions caused by burning what fuel remains.

Back in the 1970s there was very little public discussion about energy sources. In the industrialised world we had become used to the idea that electricity is generated in large centralised power stations, often out of sight as well as mind, and distributed to factories, offices, and homes by a grid system with far-reaching tentacles. Few people had any idea how the

Electricity from Sunlight   By Paul A. Lynn
© 2010 John Wiley & Sons, Ltd

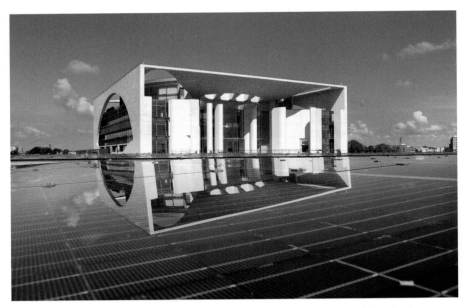

Figure 1.1 Towards the new solar age: this rooftop PV installation at the Mont-Cenis Academy in Herne, Germany, is on the site of a former coalmine (IEA-PVPS).

electricity they took for granted was produced, or that the burning of coal, oil, and gas was building up global environmental problems. Those who were aware tended to assume that the advent of nuclear power would prove a panacea; a few even claimed that nuclear electricity would be so cheap that it would not be worth metering! And university engineering courses paid scant attention to energy systems, giving their students what now seems a rather shortsighted set of priorities.

Yet even in those years there were a few brave voices suggesting that all was not well. In his famous book *Small is Beautiful*,[1] first published in 1973, E.F. Schumacher poured scorn on the idea that the problems of production in the industrialised world had been solved. Modern society, he claimed, does not experience itself as part of nature, but as an outside force seeking to dominate and conquer it. And it is the illusion of unlimited powers deriving from the undoubted successes of much of modern technology that is the root cause of our present difficulties. In particular, we are failing to distinguish between the capital and income components of the Earth's resources. We use up capital, including oil and gas reserves, as if they were steady and sustainable income. But they are actually once-and-only capital. It is like selling the family silver and going on a binge.

Schumacher's message, once ignored or derided by the majority, is increasingly seen as mainstream. For the good of Planet Earth and future generations we have started to distinguish between capital and income, and to invest heavily in renewable technologies – including solar, wind and wave power – that produce electrical energy free of carbon emissions. In recent years the message has been powerfully reinforced by former US Vice President Al Gore, whose inspirational lecture tours and video presentation *An Inconvenient Truth*[2] have been watched by many millions of people around the world.

Whereas the fossil fuels laid down by solar energy over hundreds of millions of years must surely be regarded as capital, the Sun's radiation beamed at us day by day, year by year, and century by century, is effectively free income to be used or ignored as we wish. This income is expected to flow for billions of years. Nothing is 'wasted' or exhausted if we don't use it because it is there anyway. The challenge for the future is to harness such renewable energy effectively, designing and creating efficient and hopefully inspiring machines to serve humankind without disabling the planet.

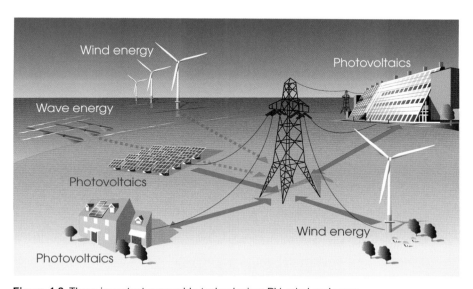

Figure 1.2 Three important renewable technologies: PV, wind and wave.

We should perhaps consider the meaning of renewable energy a little more carefully. It implies energy that is sustainable in the sense of being available in the long term without significantly depleting the Earth's capital resources, or causing environmental damage that cannot readily be repaired by nature itself. In his excellent book *A Solar Manifesto*,[3] German politician Hermann Scheer considers Planet Earth in its totality as an energy conversion system. He notes how, in its early stages, human society was itself the most efficient energy converter, using food to produce muscle power and later enhancing this with simple mechanical tools. Subsequent stages – releasing relatively large amounts of energy by burning wood; focusing energy where it is needed by building sailing ships for transport and windmills for water pumping – were still essentially renewable activities in the above sense.

What really changed things was the 19th-century development of the steam engine for factory production and steam navigation. Here, almost at a stroke, the heat energy locked in coal was converted into powerful and highly concentrated motion. The industrial society was born. And ever since we have continued burning coal, oil, and gas in ways which pay no attention to the natural rhythms of the earth and its ability to absorb wastes and byproducts, or to keep providing energy capital. Our approach has become the opposite of renewable and it is high time to change priorities.

Since the reduction of carbon emissions is a principal advantage of PV, wind, and wave technologies, we should recognise that this benefit is also proclaimed by supporters of nuclear power. But frankly they make strange bedfellows, in spite of sometimes being lumped together as 'carbon-free'. It is true that all offer electricity generation without substantial carbon emissions, but in almost every other respect they are poles apart. The renewables offer the prospect of widespread, relatively small-scale electricity generation, but nuclear must, by its very nature, continue the practice of building huge centralised power stations. PV, wind, and wave need no fuel and produce no waste in operation; the nuclear industry is beset by problems of radioactive waste disposal. On the whole renewable technologies pose no serious problems of safety or susceptibility to terrorist attack – advantages which nuclear power can hardly claim. And finally there is the issue of nuclear proliferation and the difficulty of isolating civil nuclear power from nuclear weapons production. Taken together these factors amount to a profound divergence of technological expertise and political attitudes, even of philosophy. It is not surprising that most environmentalists are unhappy with the continued development and spread of nuclear power, even though some accept that it may be hard to avoid. In part, of course, they claim that this is the result of policy failures to invest sufficiently in the benign alternatives over the past 30 or 40 years.

It would however be unfair to pretend that renewable energy is the perfect answer. For a start such renewables as PV, wind, and wave are generally diffuse and intermittent. Often, they are rather unpredictable. And although the 'fuel' is free and the waste products are minimal, up-front investment costs tend to be large. There are certainly major challenges to be faced and overcome as we move towards a solar future.

Our story now moves on towards the exciting technology of photovoltaics, arguably the most elegant and direct way of generating renewable electricity. But before getting involved in the details of solar cells and systems, it is necessary to appreciate something of the nature of solar radiation – the gift of a steady flow of energy income that promises salvation for the planet.

Figure 1.3 The promise of photovoltaics (EPIA/BP Solar).

1.2 The solar resource

The Sun sends an almost unimaginable amount of energy towards Planet Earth – around 10^{17} W (one hundred thousand million million watts). In electrical supply terms this is equivalent to the output of about one hundred million modern fossil fuel or nuclear power stations. To state it another way, the Sun provides in about an hour the present energy requirements of the entire human population for a whole year. It seems that all we need do to convert society 'from carbon to solar' is to tap into a tiny proportion of this vast potential.

However some caution is needed. The majority of solar radiation falls on the world's oceans. Some is interrupted by clouds and a lot more arrives at inconvenient times or places. Yet, even when all this is taken into account, it is clear that the Sun is an amazing benefactor. The opportunities for harnessing its energy, whether represented directly by sunlight or indirectly by wind, wave, hydropower or biomass, seem limited only by our imagination, technological skill and political determination.

The Sun's power density (i.e. the power per unit area normal to its rays) just above the Earth's atmosphere is known as the *solar constant* and equals 1366 W/m². This is reduced by around 30% as it passes through the atmos-

Figure 1.4 Energy for ever: an installation in Austria (IEA-PVPS).

phere, giving an *insolation* at the Earth's surface of about $1000\,W/m^2$ at sea level on a clear day. This value is the accepted standard for 'strong sunshine' and is widely used for testing and calibrating terrestrial PV cells and systems.

Another important quantity is the average power density received over the whole year, known as the *annual mean insolation*. A neat way of estimating it is to realise that, seen from the Sun, the Earth appears as a disk of radius R and area πR^2. But since the Earth is actually spherical with a total surface area $4\pi R^2$, the annual mean insolation just above the atmosphere must be $1366/4 = 342\,W/m^2$. However it is shared very unequally, being about $430\,W/m^2$ over the equator, but far less towards the polar regions which are angled well away from the Sun. The distribution is illustrated in the upper half of Figure 1.5.

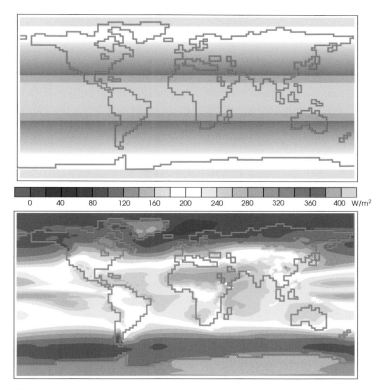

Figure 1.5 Annual mean insolation just outside the Earth's atmosphere (top) and at the Earth's surface (below). Redrawn from *Wikipedia*.

The lower half of the figure shows the reduction in insolation caused by the Earth's atmosphere. Absorption by gases and scattering by molecules and dust particles are partly responsible. Clouds are a major factor in some regions. We see that the average insolation at the Earth's surface is greatly affected by local climatic conditions, ranging from about 300 W/m^2 in the Sahara Desert and parts of the Pacific Ocean to less than 80 W/m^2 near the poles.

If we know the average insolation at a particular location, it is simple to estimate the total energy received over the course of a year (1 year = 8760 hours). For example London and Berlin, both with mean insolation of about 120 W/m^2, have annual energy totals of about $120 \times 8760/1000 = 1050$ kWh/m^2. Sydney's mean of about 200 W/m^2 is equivalent to 1750 kWh/m^2, and so on. Such figures are useful to PV system designers who need to know the total available solar resource. However, we must remember that they are averaged over day and night, summer and winter, and are likely to vary considerably from year to year. It is also interesting to speculate how far global warming, with its interruptions to historical weather patterns, may affect them in the future.

So far we have not considered the Sun's spectral distribution – that is, the range and intensity of the wavelengths in its emitted radiation. This is a very important matter because different types of solar cell respond differently to the various wavelengths in sunlight. It is well known that the Sun's spectrum is similar to that of a perfect emitter, known as a *black body*, at a temperature of about 6000 K. The smooth curve in Figure 1.6 shows that such black-body radiation spreads over wavelengths between about 0.2 and 2.0 μm, with a peak around 0.5 μm. The range of wavelengths visible to the human eye is about 0.4 μm (violet) to 0.8 μm (red). Shorter wavelengths are classed as ultraviolet (UV), longer ones as infrared (IR). Note how much of the total spectrum lies in the IR region.

The figure shows two more curves, labelled AM0 and AM1.5, representing actual solar spectral distributions arriving at Earth. To explain these we need to consider the pathlength or *Air Mass (AM)* of sunlight through the atmosphere. AM0 refers to sunlight just outside the atmosphere (pathlength zero) and is therefore relevant to PV used on Earth satellites. In the case of terrestrial PV, the pathlength is the same as the thickness of the atmosphere (AM1) when the Sun is directly overhead. But if it is not overhead the pathlength increases according to an inverse cosine law. For example when 60° from overhead the pathlength is doubled (AM2), and so on. The widely-used AM1.5 curve, shown in the figure, represents the Sun 48° from

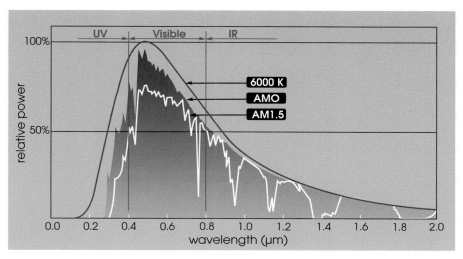

Figure 1.6 Spectral distributions of solar energy.

overhead and is generally accepted as a compromise for assessing PV cells and systems. The deep notches are due to absorption by oxygen, water vapour, and carbon dioxide.

This is not quite the whole story because when solar cells are installed at or near ground level, they generally receive indirect as well as direct solar radiation. This is shown in Figure 1.7. The *diffuse* component represents light scattered by clouds and dust particles in the atmosphere; the *albedo* component represents light reflected from the ground or objects such as trees and buildings. The electrical output from the cells depends on the combined effect of all components – direct, diffuse, and albedo. In strong sunlight the direct component is normally the greatest. But if the cells are pointed away from the Sun, or if there is a lot of cloud, the diffuse component may well dominate (clouds also cause blocking, or attenuation, of direct radiation). The albedo contribution is often small, but can be very significant in locations such as the Swiss Alps due to strong reflections from fallen snow.

We have now covered the main features of solar radiation as it affects terrestrial PV. We shall find this information useful when considering the mounting and orientation of PV cells and modules in Chapter 3.

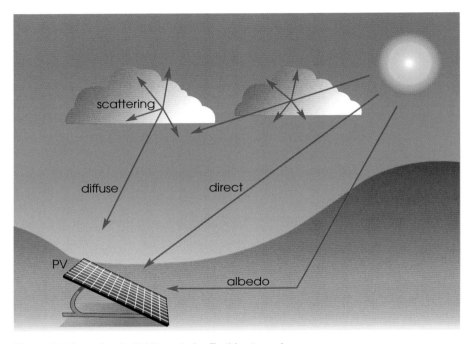

Figure 1.7 From Sun to PV through the Earth's atmosphere.

1.3 The magic of photovoltaics

From time to time human ingenuity comes up with a new technology that seems to possess a certain magic. We can all think of examples from the past – the printing press, steam locomotion, radio communication, powered flight, medical imaging – although our choices inevitably reflect personal tastes and priorities. In most cases such technologies were unimaginable to previous generations and caused amazement and even fear when they appeared. Quite often a technology that promises major social as well as commercial benefits turns out to have rather questionable applications. American aviation pioneers Wilbur and Orville Wright, whose first powered flights at Kitty Hawk in 1903 changed the world forever, initially believed that scouting aircraft would render wars obsolete by allowing each nation to see exactly what the others were doing. But by the end of World War 1 it had become clear that this view was overoptimistic and Orville instead declared that 'the aeroplane has made war so terrible that I do not believe any country will again care to start one'. Perhaps we are rather more realistic today and understand that technological advance almost always carries risk as well as social benefit. The magic is not without its downside.

Where does PV fit in the landscape of technological change? Half a century ago few people realised that sunlight could be converted directly into electricity. Even the early pioneers of PV could hardly have guessed that their researches would lead to a worldwide industry providing electricity to millions of people in developing countries. A generation ago it seemed unlikely that PV would branch out from its early success in powering space satellites and come down to Earth. More recently it would have taken courage to suggest that terrestrial PV would move into a multi-gigawatt era and start rivalling conventional methods of electricity generation in the developed world. From the technical point of view it is certainly a remarkable story – and one that, on a historical timescale, is still in its early stages.

Just as importantly, it is difficult to see any major downside in PV's gentle technology. Few people find much to object to in the deployment of solar cells and modules. True, some worry that the aesthetics of existing homes, offices, and public buildings can be marred by having PV attached to them, although this is a matter of taste. There is also the question of land use: PV takes up a lot of space compared with a conventional power plant for the same amount of electricity generation. Yet this space can often be marginal or unproductive land, in deserts or old industrial areas; and unlike wind turbines that offend many people by their visual intrusion, ground-mounted PV is hardly ever visually aggressive or unattractive. Finally there may be some risk when PV comes to the end of its useful life, but most of the materials involved are benign and the industry is very aware of its environmental credentials and the need for recycling and careful disposal. All in all the negative impacts of PV seem relatively modest, and containable.

Much of PV's magic is due to its elegance and simplicity. A solar cell turns sunlight directly into electricity without fuel, moving parts, or waste products. Made from a thin slice or layer of semiconductor material, it is literally a case of 'photons in, electrons out'. By contrast a fossil fuel or nuclear power station working on a classic thermodynamic cycle turns heat from fuel combustion or nuclear reaction into high pressure steam, then uses the steam to drive a turbine coupled to an electrical generator. This complex chain of events produces undesirable byproducts, including spent fuel and a great deal of waste heat, and in the case of fossil fuels also a lot of carbon dioxide. The high pressures and temperatures at which modern plant is

Figure 1.8 A certain magic: 'sunflowers' in Korea (IEA-PVPS).

operated put great stresses on materials and components. Small-scale electricity production using diesel generators has similar disadvantages. Meanwhile the solar cell works silently and effortlessly, a model of operational simplicity. Place it in sunlight and you can tap electricity directly from its terminals.

Not that PV is simple science. As we shall see, solar cells are high-tech products based on more than half a century of impressive research in universities, companies, and government institutes around the world. Their manufacture demands very high standards of precision and cleanliness. And they are strongly related to another modern technology that has a certain magic for many people – semiconductor electronics and computers.

1.4 A piece of history

Light has fascinated some of the world's greatest scientists. One of the most famous of them all, Isaac Newton (1642–1727), thought of it as a stream of particles, rather like miniature billiard balls. But in the early 19th century experiments by the English polymath Thomas Young and French physicist Augustin Fresnel demonstrated interference effects in light beams, which include the bands of colours often seen on the surface of soap bubbles. This suggested that light acts as a wave rather like the ripples on a pond – a theory reinforced by James Clerk Maxwell's work in the 1860s, showing visible light to be part of a very wide spectrum of electromagnetic radiation.

Yet Newton's 'billiard ball' theory refused to go away. The German physicist Max Planck used it to explain the characteristics of black-body radiation; and it subsequently proved central to Albert Einstein's famous work on the photoelectric effect in 1905, in which he proposed that light is composed of discrete miniature particles or packets of energy known as *quanta*. The subsequent development of quantum theory was one of the great intellectual triumphs of the 20th century. So our modern view is that light has an essential *duality*: for some purposes we may think of it as a stream of particles; for others, as a type of wave. The two aspects are complementary rather than contradictory.

The earliest beginnings of PV go back to 1839 when the young physicist Edmond Becquerel, working in his father's laboratory in France, discovered the photovoltaic effect as he shone light onto an electrode in an electrolyte solution. By 1877 the first solid-state PV cells had been made from selenium, and these were later developed as light meters for photography. Although a proper understanding of the phenomena was provided

Figure 1.9 Isaac Newton, Edmond Becquerel, and Albert Einstein (Wikipedia).

by quantum theory, practical application to useful PV devices had to await the arrival of semiconductor electronics in the 1950s. Thus, there was a gap of over a hundred years between Becquerel's initial discovery and the development of PV as we know it today.

The story of modern PV has been expertly reviewed in an article by solar cell pioneer Joseph Loferski,[4] formerly a professor at Brown University in the USA. Although we must leave aside the technical details of his account, the following broader points will serve well to bring the PV story up towards the end of the 20th century.

The modern PV age may be said to have begun in 1954 with the work of researchers at the Bell Telephone and RCA laboratories, who reported new types of semiconductor devices, based on silicon and germanium, that were an order of magnitude more efficient than previous cells at converting radiation directly into electricity. The fledgling PV community hoped that this would lead to new applications for solar cells including electrical power generation. However their hopes were not realised, in part because that decade was a time of great expectations for nuclear energy. Sceptics believed that solar energy was too diffuse and intermittent, and the new devices far too expensive. At that moment in history PV looked rather like a solution in search of a problem.

What changed the situation almost overnight was the launch of the first earth satellite, the USSR's *Sputnik*, in 1957. Satellites and solar cells – even expensive ones – were made for each other. The early satellites needed only a very modest amount of electricity, and the weight and area of solar panels needed to produce this were acceptable to satellite designers. Also, the types of cell made in 1954 were proving reliable and seemed likely to operate in the space environment for many years without significant deterioration. The first US research satellite using a PV power supply was launched in 1958. It was the size of a large grapefruit. Its solar cells covered an area of about $100 \, cm^2$ and produced just a few tens of milliwatts. In 1962 the first-ever commercial telecommunication satellite, *Telstar*, was launched with sufficient solar cells to produce 14 W from the Sun. By the early 1970s, space satellites powered by solar cells had become quite commonplace. PV in space had already made its mark.

The possibilities for 'bringing PV down to Earth' depended crucially on lowering the price of solar cells. In 1970 the US price was around $300 per peak watt (we normally quote solar cell power in peak watts (W_p), being the rated power at a standard insolation of $1000 \, W/m^2$). This was acceptable for extremely expensive space satellites, but hopeless for terrestrial electricity production on a significant scale. What encouraged PV researchers to

Figure 1.10 Telstar (Wikipedia/ NASA).

hold on to their dream was the realisation that prices would almost certainly fall dramatically as production levels rose, in accordance with the well-known 'learning curve' concept. Experience had shown that, for every doubling of cumulative production of a wide range of manufactured products, price tended to drop between 10 and 30%. For mature technologies such as steel or electric motors, such doubling, given the high current production levels, would require many decades. But a fledgling technology like PV had many doublings of cumulative production to look forward to, so major reductions in cost could be expected over a comparatively short timescale. Indeed, it was predicted that by the time cumulative PV cell production reached gigawatt levels (1 gigawatt (GW) being equal to 1000 megawatts), as required for serious terrestrial application, the price would have dropped nearer to $1/W_p$. As with other technologies, new inventions and manufacturing systems that could not be visualised would arise, unpredictable political and economic factors would occur – and the price would be driven down.

This apparently bold prognosis, which helped lift the gloom and spurred the PV community on to ever greater efforts, proved farsighted. The actual 'learning curve' for world PV production over the period 1987 – 2009 is shown in Figure 1.11, plotted on logarithmic scales in terms of euros per

15

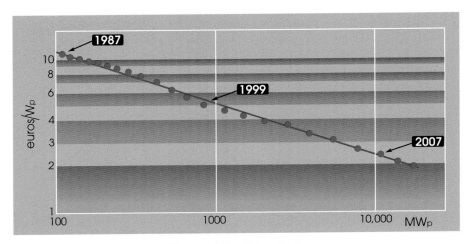

Figure 1.11 The 'learning curve' for cumulative PV production.

peak watt (euros/W_p) against cumulative peak megawatts (MW_p). We see that as cumulative production advanced from 100 MW_p in 1987 to 1000 MW_p in 1999, the cost per peak watt fell from about €11 to €5. Another tenfold advance to 10 000 MW_p was achieved by 2007 and the cost reduced to around €2.5. This excellent trend corresponds to an average cost reduction of about 20% for every doubling of cumulative production – much as expected for manufactured products. The growth has continued since 2007, leading experts to predict that PV will become truly competitive with conventional electricity generation in much of the developed world in the coming decade. These are exciting times!

Returning now to our historical review, the renewed optimism of the PV community, based on 'learning curve' predictions, was bolstered by the first 'oil shock' in 1973 when oil-producing countries decided greatly to increase the price of crude oil and exert more control over its supply. Funding for PV research and development in the USA under President Jimmy Carter then increased dramatically. Unfortunately, government support was subsequently cut back hard by President Ronald Reagan's administration from 1980 onwards, but great advances had already been made, and major PV research programmes in Germany and Japan were adding their own important contributions. The efficiencies of solar cells were constantly being raised: new PV materials and cell structures were being investigated; and on the applications front, a range of PV power plants emerged with megawatt capacity. In 1985 Professor Martin Green's group at the University of New South Wales in Australia pushed the efficiency of new-design silicon cells above the 20% barrier – some four times higher than the cells that had heralded the arrival of the modern PV age in 1954. By 1990 the same group achieved efficiencies of 24% and it has continued as a major pacesetter for crystalline silicon cells, the long-term workhorse of the PV industry. From such developments was the 'learning curve' constructed.

The pace of PV research, development, and application continues unabated today. Growing awareness of global warming and the vital role of the renewable energies in combating it have ensured that governments around the world appreciate the need to encourage and stimulate PV, and we are now in the multi-gigawatt era. This must have been almost unimaginable half a century ago.

At the end of his 1993 article Joseph Loferski noted that the blossoming PV edifice was destined to grow many-fold again. The small band of researchers who had ushered in the modern PV age in the 1950s had multiplied into 'a band of brothers and sisters, we happy few', who shared the dream that solar PV electricity was destined for an ever-greater future in the service of humanity. From today's perspective, his vision and optimism seem entirely justified.

1.5 Coming up to date

How can we summarise the current status of a technology such as PV that has been, and still is, experiencing dramatic growth? Today's research and development, novel PV installations, and global statistics will very soon seem history. But fortunately certain trends that have developed over the past 15 or 20 years seem likely pointers to the future. We can discuss these trends more easily by dividing PV systems into two broad categories: *grid-connected systems* (also called *grid-tied systems*) that feed any surplus PV electricity into a grid and accept electricity from the grid when there is a solar deficit; and *stand-alone systems* that are self-contained and not tied to a conventional electricity grid. These categories may usefully be subdivided as follows:

Grid-connected systems

    1. Building-integrated photovoltaics (BIPV) on roofs or facades of individual houses, offices, factories and other commercial premises

(or on adjacent land) – covering a wide power range from about $1\,kW_p$ to several MW_p.

2. Land-based PV power plants, often remote from individual electricity consumers – typically 1–$20\,MW_p$, with a few up to $50\,MW_p$, and much larger ones planned.

Stand-alone systems

3. Low-power solar home systems (SHSs), supplying small amounts of electricity to individual homes in developing countries – typically 30–$100\,W_p$.

4. Higher-power systems for isolated homes and buildings in the developed world – typically 1–$20\,kW_p$.

5. PV systems for a wide range of applications, including water pumping and irrigation, isolated telecommunications equipment, marine buoys, traffic control signs, and solar-powered cars and boats.

In the 1980s PV started to make a major contribution, supplying small amounts of electricity to the millions of families in 'sunshine countries' of the developing world with no access to, or promise of, an electricity grid. This was rightly seen as a noble social objective as well as a commercial opportunity that would increase PV's international market. However it became increasingly evident that solar home systems (item 3 above) could not, by themselves, boost global production towards the levels dreamed of by the PV community. A typical SHS requires only one small PV panel, but even so the 'up-front' costs cannot easily be afforded by individual families in developing countries without effective government financing schemes that are not always forthcoming. And maintenance problems (generally with batteries or other system components rather than the PV itself) can easily reduce reliability. So although the SHS market is socially important and continues to grow, it no longer represents a major plank of the global PV industry.

Various other types of stand-alone system were steadily developed in the 1980s, often providing valuable PV electricity in remote locations that would otherwise need diesel generators. In addition a number of grid-connected PV power stations were commissioned, mainly in the USA, by electric utilities keen to assess the commercial possibilities and reliability of the new technology. However, the limited number and scale of all these systems offered little prospect for the exciting expansion of PV needed to make it a major source of electricity.

Figure 1.12 This PV module powers a solar home system in Bolivia (EPIA/BP Solar).

What really changed the outlook for global PV production was the emphatic shift towards grid-connected systems in the developed world that got under way in the 1990s. It was the citizens of richer countries that would provide the up-front costs and market stimulus to propel PV faster along its 'learning curve', leading to price reductions as cumulative production really took off. This policy shift was supported by increasing awareness among governments of the importance of renewable energy for combating climate change, and by the growing enthusiasm of individuals and companies to 'do their bit' by installing BIPV systems, even though the price of solar electricity was not yet competitive. Electricity utilities began to accept that the flow of electricity was not all 'one-way', allowing customers to be providers as well as consumers, and introducing tariffs – often not very generous ones – for feeding electricity back into the grid.

As far as governments are concerned, the price support mechanisms devised for grid-connected systems have proved crucial. PV is similar to other

Figure 1.13 A 9 kW grid-connected PV system in Northern Italy (IEA-PVPS).

renewables such as hydroelectric and wind in having high up-front capital costs and very low running costs. But this can make it hard for families and organisations to find the initial capital, and even harder if they are not guaranteed an attractive price for surplus PV electricity fed back into the grid. In recent years many governments have provided capital grants to encourage people to install domestic PV systems; and the more far-sighted ones have introduced generous *feed-in tariffs* that offer long-term, guaranteed, payments for renewable electricity. Countries that have given PV a big boost with feed-in tariffs – especially Germany and Spain – have stimulated their home markets and, by doing so, have pushed PV decisively along its 'learning curve' into the multi-gigawatt era. As cumulative world production surges and the price comes down, poorer families in sunshine countries are more likely to get their solar home systems.

Not that feed-in tariffs and other forms of government support are universally popular. Some politicians tend to regard them with suspicion, arguing that market forces alone should determine the price of PV and other renewables. If PV is currently too expensive it should be left to develop and mature, finding its own level of support. Others are more likely to vote for taxpayers' or consumers' money being used to support a new and promising technology that will, in due course, benefit the whole of society as well as the planet. Such differences tend to produce stop–go support for PV when governments change, causing confusion and clouding investment decisions. Yet in spite of these drawbacks, the clear trend is towards support by governments regardless of their political hue, mainly because of near-universal agreement that global warming must be checked and renewable energy championed.

The emphasis on grid-connected systems in the developed world continues today, making them far more important than stand-alone systems in terms of the total volumes of PV required. Huge numbers of rooftop installations are being installed on homes; offices and commercial buildings increasingly use PV on their roofs and facades; and large factory rooftops are being fitted with PV, sometimes retrospectively. The market for power plants is also developing rapidly, with Germany, Spain and the USA prominent in

Figure 1.14 115 kWp rooftop installation of the Ford Motor Company in London (IEA-PVPS).

21

pressing ahead with ever larger installations. In 2009 it was announced that a huge plant is to be built in a remote desert region of Mongolia, working up to a total capacity of $2\,GW_p$ over a ten year period.

So where has all this activity got us? We previously noted that world cumulative PV production passed the $10\,GW_p$ landmark in 2007. The industry is now in an exciting new phase, with multi-gigawatt annual production set to challenge fossil fuel and nuclear plants and achieve 'grid-parity'. Crystal ball gazing is always risky, but if current and projected increases in cumulative production are maintained it seems possible that we will be approaching $1000\,GW_p$ of PV installed around the world by 2020 or soon after. Recalling that a large conventional power station generates about 1 GW, it is clear that renewable electricity on this scale would make a serious contribution to global supplies.

This raises an interesting question: what total area will be required to accommodate all this PV? After all, sunlight is not a highly concentrated energy source and $1000\,GW_p$ of installed capacity would take up a large area. Will Planet Earth be smothered with solar cells? An approximate answer may be found by noting that $1\,kW_p$ of solar modules takes up a typical area of about $10\,m^2$. However, modules cannot generally be crammed together, especially in large installations where space may be needed to allow servicing or prevent shading, so we might allow $20\,m^2$ per kW_p. This means that $1000\,GW_p$, equal to $1000 \times 10^6\,kW_p$, would need around $20000 \times 10^6\,m^2$, which could be provided by a $140 \times 140\,km$ square of land, roughly three times the area taken up by London and its suburbs, or by Paris. In other words our projected PV scenario for 2020 might require a total area comparable with three large modern cities – but spread right around the globe. When we consider that huge arid regions and deserts of the world, marginal and ex-industrial land, and hundreds of millions of rooftops on houses and commercial buildings are all candidates for PV, there appears to be plenty of space!

Our brief summary of recent developments and likely trends has so far ignored one of the most important aspects – research and development (R&D) of solar cells, modules, and the additional items that go to make up a complete PV system. In fact the past 20 years has seen extraordinary R&D activity by teams in universities, government institutes, and PV companies. Solar cells are constantly being improved, new types of cell invented, and system components improved in reliability as well as reduced in price. However we will be in a better position to consider such topics after covering some of the basic science of solar cells in the next chapter.

Figure 1.15 PV power plant in Colorado, USA (IEA-PVPS).

References

1. E.F. Schumacher. *Small is Beautiful*, 1st edn 1973, republished by Hartley & Marks: London (1999).
2. Al Gore. *An Inconvenient Truth*, Bloomsbury Publishing: London (2006).
3. Hermann Scheer. *A Solar Manifesto*, James & James: London (2005).
4. J.J. Loferski. The first forty years: a brief history of the modern photovoltaic age. *Progress in Photovoltaics: Research and Applications*, **1**, 67–78 (1993).

2 Solar cells

2.1 Setting the scene

We are now ready to discuss the underlying principles and operation of the invention central to our story – the modern solar cell. To help set the scene we shall also say a few words about PV modules, reserving detailed discussion to the next chapter. It will be helpful to start this chapter with a brief account of the main types of solar cell and module in widespread use today.

Silicon solar cells have been the workhorse of the PV industry for many years and currently account for well over 80% of world production. Modules based on these cells have a long history of rugged reliability, with guarantees lasting 20 or 25 years that are exceptional among manufactured products. Although cells made from other materials are constantly being developed and some are in commercial production, it will be hard to dislodge silicon from its pedestal. The underlying technology is that of semiconductor electronics: a silicon solar cell is a special form of semiconductor diode. Fortunately, silicon in the form of silicon dioxide (quartz sand) is an extremely common component of the Earth's crust and is essentially non-toxic. There is a further good reason for focussing strongly on silicon cells in this chapter: in its *crystalline* form silicon has a simple lattice structure, making it comparatively easy to describe and appreciate the underlying science.

There are two major types of crystalline silicon solar cell in current high-volume production:

Electricity from Sunlight   By Paul A. Lynn
© 2010 John Wiley & Sons, Ltd

Figure 2.1 Each of these PV modules contains 72 monocrystalline silicon solar cells (EPIA/ Phoenix Sonnenstrom).

- *Monocrystalline.* The most efficient type, made from a very thin slice, or wafer, of a large single crystal obtained from pure molten silicon. The circular wafers, often 5 or 6 inches (15 cm) in diameter, have a smooth silvery appearance and are normally trimmed to a pseudo-square or hexagonal shape so that more can be fitted into a module – see Figure 2.1. Fine contact fingers and busbars are used to conduct the electric current away from the cells which have a highly ordered crystal structure with uniform, predictable, properties. However, they require careful and expensive manufacturing processes, including 'doping' with small amounts of other elements to produce the required electrical characteristics. Typical commercial module efficiencies fall in the range 12–16%. The module surface area required is about $7 \, \text{m}^2/\text{kW}_\text{p}$.

- *Multicrystalline,* also called *polycrystalline.* This type of cell is also produced from pure molten silicon, but using a casting process. As the silicon cools it sets as a large irregular multicrystal which is then cut into thin square or rectangular slices to make individual cells. Their crystal structure, being random, is less ideal than with

Figure 2.2 The façade of this cable-car station in the Swiss Alps is covered with multicrystalline silicon PV modules (IEA-PVPS).

monocrystalline material and gives slightly lower cell efficiencies, but this disadvantage is offset by lower wafer costs. Cells and modules of this type often look distinctly blue, with a scaly, shimmering appearance, as in the building façade shown in Figure 2.2. Multicrystalline modules exhibit typical efficiencies in the range 11–15% and have overtaken their monocrystalline cousins in volume production over recent years. The module surface area is about $8\,\mathrm{m^2}\,/\mathrm{kW_p}$.

You have probably already gathered that the *efficiency* of any solar cell or module, the percentage of solar radiation it converts into electricity, is considered one of its most important properties. The higher the efficiency, the smaller the surface area for a given power rating. This is important when space is limited, and also because some of the additional costs of PV systems – especially mounting and fixing modules – are area related. Crystalline silicon cells, when operated in strong sunlight, have the highest

27

efficiencies of all cells commonly used in terrestrial PV systems, plus the promise of modest increases as the years go by due to improvements in design and manufacture. But it is important to realise that other types of cell often perform better in weak or diffuse light, a matter we shall return to in later sections.

Research laboratory cells achieve considerably higher efficiencies than mass-produced cells. This reflects the ongoing R&D effort that is continually improving cell design and leading to better commercial products. In some applications where space is limited and efficiency is paramount – for example, the famous solar car races held in Australia – high-quality cells made in small batches are often individually tested for efficiency before assembly.

Module efficiencies are slightly lower than cell efficiencies because a module's surface area cannot be completely filled with cells and the frame also takes up space. It is always important to distinguish carefully between cell and module efficiency.

There is one further type of silicon solar cell in common use:

- ▪ *Amorphous.* Most people have met small amorphous silicon (a-Si) cells in solar-powered consumer products such as watches and calculators that were first introduced in the 1980s. Amorphous cells are cheaper than crystalline silicon cells, but have much lower efficiencies, typically 6–8%. Nowadays, large modules are available and suitable for applications where space is not at a premium, for example on building facades. The surface area required is about $16\,m^2/kW_p$. We shall discuss amorphous silicon in Section 2.3.

We focus initially on crystalline silicon solar cells for two main reasons: their comparatively simple crystal structure and theoretical background; and their present dominant position in the terrestrial PV market. Their wafer technology has been around for a long time and is often referred to as 'first generation'; they are the cells you are most likely to see on houses, factories, and commercial buildings.

However, it is important to realise that many other semiconductor materials can be used to make solar cells. Most come under the heading of *thin-film* – somewhat confusing because a-Si is also commonly given this title – and involve depositing very thin layers of semiconductor on a variety of substrates. Thin-film products are generally regarded as the ultimate goal for terrestrial PV since they use very small amounts of semiconductor material and large-scale continuous production processes without any need to cut and mount individual crystalline wafers. Thin-film modules based on the compound semiconductors *copper indium diselenide (CIS)* and *cadmium telluride (CdTe)* are in commercial production. Often referred to as 'second-generation', they currently have efficiencies lower than those of crystalline silicon, but they represent a highly significant advance into thin-film products. We will discuss them, and several types of specialised cells and modules, later in the chapter.

2.2 Crystalline silicon

2.2.1 The ideal crystal

A large single crystal of pure silicon forms the starting point for the mono-crystalline silicon solar cell – the most efficient type in common use. As we shall see, the simple and elegant structure of such crystals makes it comparatively easy to explain the basic semiconductor physics and operation of PV cells. We are talking here of silicon refined to very high purity, similar to that used by the electronics industry to make semiconductor devices (diodes, transistors, and integrated circuits including computer chips). Its purity is typically 99.99999%. This contrasts with the far less pure metallurgical grade silicon, produced by reducing quartzite in electric arc furnaces, that is used to make special steels and alloys.

The *Czochralski (CZ)* method of growing silicon crystals is quite easy to visualise. Chunks of pure silicon with no particular crystallographic structure are melted at 1414 °C in a graphite crucible. A small seed of silicon is then brought into contact with the surface of the melt to start crystallisation. Molten silicon solidifies at the interface between seed and melt as the seed is slowly withdrawn. A large ingot begins to grow both vertically and laterally with the atoms tending to arrange themselves in a perfect crystal lattice.

Unfortunately, this classic method of producing crystals has a number of disadvantages. Crystal growth is slow and energy intensive, leading to high production costs. Impurities may be introduced due to interaction between the melt and the crucible. And in the case of PV the aim is of course to produce thin solar cell wafers rather than large ingots, so wire saws are used to cut the ingot into thin slices, a time-consuming process that involves discarding valuable material. For these reasons the PV industry has spent a lot of R&D effort investigating alternatives, including pulling crystals in thin sheet or ribbon form, and some of these are now used in volume

Figure 2.3 Chunks of silicon (EPIA/Photowatt).

production. Whatever method is employed, the desired result is pure crystalline silicon with a simple and consistent atomic structure.

The element silicon has atomic number 14, meaning that each atom has 14 negatively charged electrons orbiting a positively charged nucleus, rather like a miniature solar system. Ten of the electrons are tightly bound to the nucleus and play no further part in the PV story, but the other four *valence electrons* are crucial and explain why each atom aligns itself with four immediate neighbours in the crystal. This is illustrated by Figure 2.4(a). The 'glue' bonding two atoms together is two shared valence electrons, one from each atom. Since each atom has four valence electrons that are not tightly bound to its nucleus, a perfect lattice structure is formed when each atom forms bonds with its four nearest neighbours (which are actually at the vertices of a three-dimensional tetrahedron, but shown here in two dimensions for simplicity). The structure has profound implications for the fundamental physics of silicon solar cells.

Silicon in its pure state is referred to as an *intrinsic* semiconductor. It is neither an insulator like glass, nor a conductor like copper, but something in between. At low temperatures its valence electrons are tightly con-

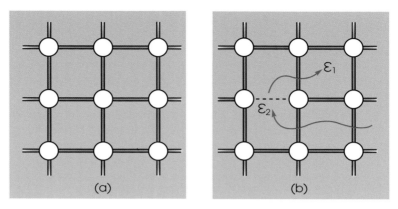

Figure 2.4 (a) Silicon crystal lattice; (b) electrons and holes.

strained by bonds, as in part (a) of the figure, and it acts as an insulator. But bonds can be broken if sufficiently jolted by an external source of energy such as heat or light, creating electrons that are free to migrate through the lattice. If we shine light on the crystal the tiny packets, or *quanta*, of light energy can produce broken bonds if sufficiently energetic. The silicon becomes a conductor, and the more bonds are broken the greater its conductivity.

Figure 2.4(b) shows an electron ε_1 that has broken free to wander through the lattice. It leaves behind a broken bond, indicated by a dotted line. The free electron carries a negative charge and, since the crystal remains electrically neutral, the broken bond must be left with a positive charge. In effect it is a positively charged particle, known as a *hole*. We see that breaking a bond has given rise to a pair of equal and opposite charged 'particles', an electron and a hole. Not surprisingly they are referred to as an *electron–hole pair*.

At first sight the hole might appear to be an 'immovable object' fixed in the crystal lattice. But now consider the electron ε_2 shown in the figure, which has broken free from somewhere else in the lattice. It is quite likely to jump into the vacant spot left by the first electron, restoring the original broken bond, but leaving a new broken bond behind. In this way a broken bond, or hole, can also move through the crystal, but as a positive charge. It is analogous to a bubble moving in a liquid; as the liquid moves one way the bubble is seen travelling in the opposite direction.

We see that the electrical properties of intrinsic silicon depend on the number of mobile electron–hole pairs in the crystal lattice. At low temperatures, in the dark, it is effectively an insulator. At higher temperatures, or in sunlight, it becomes a conductor. If we attach two contacts and apply an external voltage using a battery, current will flow – due to free electrons moving one way, holes the other. We have now reached an important stage in understanding how a silicon wafer can be turned into a practical solar cell.

Yet there is a vital missing link: remove the external voltage and the electrons and holes wander randomly in the crystal lattice with no preferred directions. There is no tendency for them to produce current flow in an external circuit. A pure silicon wafer, even in strong sunlight, cannot *generate* electricity and become a solar cell. What is needed is a mechanism to propel electrons and holes in opposite directions in the crystal lattice, forcing current through an external circuit and producing useful power. This mechanism is provided by one of the great inventions of the 20th century, the semiconductor *p–n* junction.

2.2.2 The *p–n* junction

A conventional monocrystalline solar cell has a silvery top surface surmounted by a fine grid of metallic fingers forming one of its electrical contacts. What is less obvious is that the cell actually consists of two different layers of silicon that have been deliberately *doped* with very small quantities of impurity atoms, often phosphorus and boron, to form a *p–n junction*. The addition of such *dopants* is absolutely crucial to the cell's operation and provides the mechanism which forces electrons and holes generated by sunlight to do useful work in an external circuit.

The *p–n* junction may be regarded as the basic building block of the semiconductor revolution that began back in the 1950s. It is perhaps a little surprising that an invention normally associated with mainstream electronics should also form the basis of PV technology; but a silicon solar cell is essentially a form of *p–n* junction specially tailored to the task of converting sunlight into electricity.

We have already noted that heating or shining light on pure silicon can alter its electrical properties, progressively converting it from an insulator into a conductor. Another extremely important way of modifying its properties is to add small amounts of dopants. For example if phophorus is added to molten silicon, the solidified crystal contains some phosphorus atoms in place of silicon. While the latter has four valence electrons able to form bonds with neighbouring atoms, phosphorus has five. The extra one is only weakly bound to its parent atom and can easily be enticed away, as shown

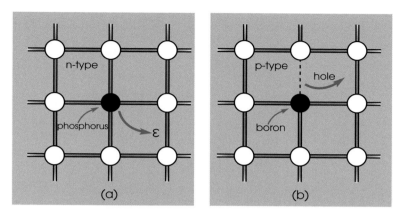

Figure 2.5 (a) A phosphorus atom in *n*-type silicon provides an extra free electron; (b) a boron atom in *p*-type silicon provides an extra hole.

in Figure 2.5(a). In other words silicon doped with phosphorus provides plenty of free electrons, known as the *majority carriers*. Generally there are also a few holes present due to thermal generation of electron–hole pairs, as in intrinsic silicon, and these are called *minority carriers*. The material is a fairly good conductor and is referred to as negative-type or *n-type*.

A complementary situation arises if silicon is doped with boron, which has only three valence electrons loosely bound to its nucleus, illustrated in part (b) of the figure. Each boron atom can only form full bonds with three neighbouring silicon atoms, so boron introduces broken bonds into the crystal. In this case holes are the majority carriers and electrons the minority carriers. Once again, the material becomes a conductor; it is referred to as positive-type or *p-type*.

We see that *n*-type material has many surplus electrons and *p*-type material has many surplus holes. The next step is to consider what happens when the two materials are joined together to form a *p–n* junction, illustrated in Figure 2.6(a).

Near the interface, free electrons in the *n*-type material start diffusing into the *p*-side, leaving behind a layer that is positively charged due to the presence of fixed phosphorus atoms. Holes in the *p*-type material diffuse into the *n*-side, leaving behind a layer that is negatively charged by the fixed boron atoms. This diffusion of the two types of majority carriers, in opposite directions across the interface, has the extremely important effect

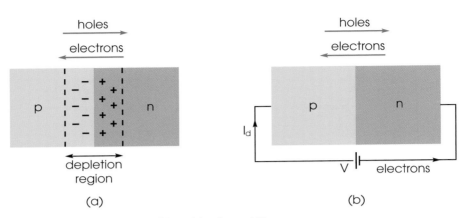

Figure 2.6 (a) A *p–n* junction; (b) applying forward bias.

of setting up a strong electric field, creating a potential barrier to further flow. Equilibrium is established when the tendency of electrons and holes to continue diffusing down their respective concentration gradients is offset by their difficulty in surmounting the potential barrier. In this condition there are hardly any mobile charge carriers left close to the junction and a so-called *depletion region* is formed.

The depletion region makes the *p–n* junction into a diode, a device that conducts current easily in one direction only. Figure 2.6(b) shows an external voltage *V* applied to the diode, making the *p*-type material positive with respect to the *n*-type, referred to as *forward bias*. In effect the external voltage counteracts the 'built-in' potential barrier, reducing its height and encouraging large numbers of majority carriers to cross the junction – electrons from the *n*-side and holes from the *p*-side. This results in substantial forward current flow (note that conventional positive current is actually composed of negatively charged electrons flowing the other way; we may think of them as going right round the circuit through the battery and back into the *n*-type layer). Conversely if the external voltage is inverted to produce a *reverse bias,* the potential barrier increases and the only current flow is a very small *dark saturation current (I_0)*. This is because a bias that increases the potential barrier for majority carriers decreases it for minority carriers – and at normal temperatures there are some of these present on both sides of the junction due to thermal generation of electron–hole pairs.

The practical result of these movements of electrons and holes is summarised by the diode characteristic in Figure 2.7. Diode current *I* increases

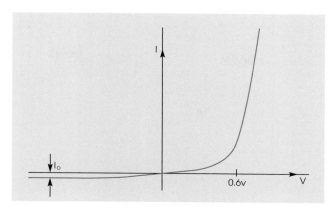

Figure 2.7 The voltage–current characteristic of a silicon diode.

with positive bias, growing rapidly above about 0.6 V; but with negative bias the reverse current 'saturates' at a very small value I_0. Clearly this device only allows current flow easily in one direction. Mathematically the curve is expressed as:

$$I = I_0\left[\exp\left(\frac{qV}{kT}\right) - 1\right] \tag{2.1}$$

where q is the charge on an electron, k is Boltzmann's constant, and T is the absolute temperature.

You are perhaps beginning to wonder what all this has to do with solar cells, because we have not so far discussed the effects of shining light on the diode and it is not obvious what these will be. However, rest assured that understanding the above discussion of electrons and holes, majority and minority carriers, and potential barriers is essential for unravelling the mysteries of photovoltaics!

2.2.3 Monocrystalline silicon

2.2.3.1 Photons in action

We are now close to understanding how a monocrystalline silicon wafer, doped to create a semiconductor diode, can work as a power-generating solar cell. The basic scheme of Figure 2.8 shows a small portion of such a cell. At the top several metallic contact fingers form part of the cell's negative terminal. Next comes a thin layer of n-type material interfacing with

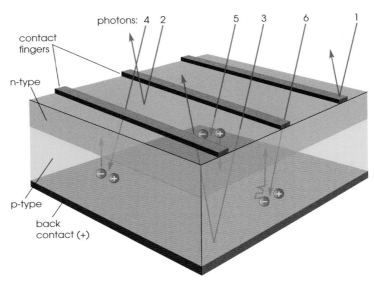

Figure 2.8 The basic scheme of a crystalline silicon solar cell.

a thicker layer of *p*-type material to produce the crucial *p–n* junction. And finally there is a back contact that acts as the positive terminal. For clarity the cell's thickness is exaggerated in the figure; it is actually a very slim wafer, normally less than 0.3 mm from top to bottom.

A stream of photons containing minute packets or *quanta* of energy shines on the cell. Their numbers are staggering: in strong sunlight a 6 inch (15 cm) cell receives more than 10^{19} photons every second. Various possible fates await them, some productive, others fruitless, and we show a few important examples in the figure.

Unfortunately, there is some loss of photons by optical reflection back from the conducting fingers, top surface, and rear surface (nos. 1, 2, and 3 in the figure). The rest enter the cell body, but only those with a certain minimum energy, known as the *bandgap*, have any chance of creating an electron–hole pair and contributing to the cell's electrical output. The most productive ones, for reasons explained below, create electron–hole pairs in the *n*-type layer or in the *p*-type layer very close to the junction (4 and 5). Less productive, on average, are the ones that travel further into the *p*-type material (6). Successful cell design involves producing as many electron–hole pairs as possible, preferably close to the junction. But even high-quality cells are subject to theoretical limits dictated by the spectral distribution of sunlight, the nature of light absorption in silicon, and quantum theory. We shall discuss these topics a little later.

First comes the big question: what happens to the electron–hole pairs generated within the cell by sunlight, and how do they produce current flow in an external circuit?

As we have seen, majority carriers (electrons in n-type material, holes in p-type) are the main players in a conventional semiconductor diode. By initial diffusion across the p–n junction they set up a depletion layer and create a potential barrier. Forward-biasing the diode reduces the height of the barrier, making it easier for them to cross the junction and produce substantial current. In reverse bias the barrier increases and current flow is severely inhibited. Diode action is principally due to the behaviour of majority carriers under the influence of an applied external voltage.

With solar cells, however, it is light-generated minority carriers that take centre stage. The basic reason may be simply stated: a potential barrier that inhibits transfer of majority carriers across a p–n junction positively encourages the transfer of minority carriers. Whereas majority carriers experience 'a hill to climb', minority carriers see 'a hill to roll down'. With luck they are swept down this hill, *collected* at the cell terminals, and produce an output current proportional to the intensity of the incident light.

Let us consider the three photons in Figure 2.8 that successfully create electron–hole pairs in the crystal lattice. Number 4 produces a pair in the p-type region, close to the junction. Its free electron, a minority carrier in p-type material, is easily swept across the junction and collected. So is the hole produced in the n-type region by number 5, which is swept across the junction in the opposite direction. Both these minority carriers should contribute to the light-generated current.

Photon 6 also creates an electron–hole pair, but well away from the junction and its associated electric field. The free electron does not immediately experience 'a hill to roll down', but instead starts wandering randomly through the silicon lattice. In the figure it is shown eventually reaching the junction and being swept away to success. But the journey is a dangerous one: it may instead encounter a hole and be annihilated. Although such *recombination* is not illustrated in the figure, unfortunately it occurs not only in the main body of the cell (*bulk recombination*) but even more importantly at the edges and metal contacts due to defects and impurities in the crystal.

The longer a minority carrier wanders around, the greater the distance travelled through the crystal and the more likely it is to be lost by recombination. Two measures are used to describe the risk. The *carrier lifetime* is the average amount of time between electron–hole generation and recombination (the bigger the better) which for silicon is typically $1\,\mu s$. The *diffusion length* is the average distance a carrier moves from the point of generation until it recombines, for silicon typically $0.2\,mm$ which is comparable with the thickness of the monocrystalline wafer. This again emphasises the value of electron–hole pairs generated close to the junction.

We have now covered some fundamental aspects of solar cell operation, including the key role played by light-generated minority carriers. The next task is to consider the voltage–current characteristics of the cell as measured at its output terminals.

2.2.3.2 Generating power

We have seen solar photons at work, creating minority carriers that speed towards the solar cell's output terminals under the magical influence of the *p–n* junction. But how is all this internal activity reflected in the cell's power generation, and what voltages and currents are produced at its terminals? Figure 2.9(a) helps answer the question with an equivalent circuit summarising the cell's behaviour as a circuit component. It consists of a diode representing the action of the *p–n* junction together with a current generator representing the light-generated current I_L.

In dark conditions I_L is zero and the cell is quiescent. If an external voltage source is connected the cell behaves just like a semiconductor diode with the characteristic shown in part (b) of the figure (this has the same form as Figure 2.7). We choose to define the current I as flowing into the circuit and, in the dark, it must be the same as the diode current I_D. Note also that since a diode is a *passive* device that dissipates power, the cell's dark characteristic lies entirely in the first and third quadrants (I and V both positive, or both negative). But if sufficient sunlight falls on the cell to turn it into an *active* device delivering power to the outside world, the current I must reverse and the characteristic will shift into the fourth quadrant (I negative, V positive) shown shaded in the figure.

In sunlight the generator produces a current I_L proportional to the level of insolation. It is effectively superimposed on the normal diode characteristic, and we may write:

$$I = I_D - I_L \tag{2.2}$$

Substituting for the diode current using Equation (2.1) gives:

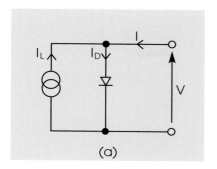
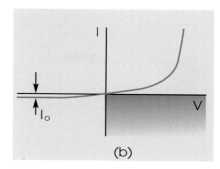

Figure 2.9 (a) The equivalent circuit of a solar cell; (b) its *I–V* characteristic in the dark.

$$I = I_0\left[\exp\left(\frac{qV}{kT}\right) - 1\right] - I_L \qquad (2.3)$$

This equation confirms that the diode *I–V* characteristic is shifted down into the fourth quadrant by an amount equal to the light-generated current I_L. This is shown in Figure 2.10(a).

Most people are unfamiliar with curves in the fourth quadrant, so for convenience the *I–V* characteristics of a solar cell are normally 'flipped over' to the first quadrant. This is equivalent to plotting *V* against *–I*. Part (b) of the figure illustrates a family of such curves for a typical crystalline silicon cell rated by the manufacturer at 2 W_p. Each curve represents a different strength of sunlight, and hence a different value of I_L . You will recall that PV cells and modules are normally rated in peak watts (W_p), indicating the maximum power they can deliver under standard conditions (insolation 1000 W/m^2, cell temperature 25 °C, AM1.5 solar spectrum). Therefore we should first consider how the rated power of 2 W_p relates to the 1000 W/m^2 *I–V* curve.

In general the cell's power output equals the product of its voltage and current. No power is produced on open circuit (maximum voltage, zero current) or short circuit (maximum current, zero voltage). The full rated power is obtained by operating the cell slightly below maximum voltage and current at its *maximum power point* (*MPP*), shown as P_1 against the 1000 W/m^2 curve, and corresponding to about 4 A at 0.5 V, or 2 W. We can only obtain the promised output power by operating the cell at its MPP. Three other curves are shown for lower insolation values of 750, 500 and 250 W/m^2; each has its own MPP(P_2, P_3, P_4) indicating the maximum power available from the cell at that particular strength of sunlight.

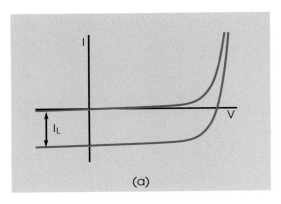

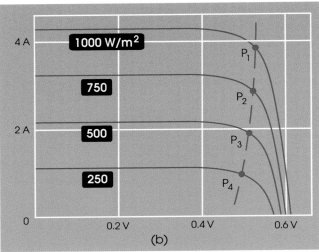

Figure 2.10 (a) The light-generated current shifts the cell's characteristic into the fourth quadrant; (b) a family of *I–V* curves for a 2 W$_p$ solar cell.

Note that the maximum voltage produced by a silicon solar cell is about 0.6 V, considerably less than the 1.5 V of a dry battery cell. This means that it is essentially a low-voltage, high-current, device and many cells must be connected in series to provide the higher voltages required for most applications. For example the PV module previously illustrated in Figure 2.1 has 72 individual cells connected in series, giving a DC voltage of about 35 V at the MPP. Higher voltages may be obtained by connecting a number of modules in series.

The *I–V* characteristics suggest another important aspect of the solar cell – it is helpful to think of it as a *current source* rather than a *voltage source* like a battery. A battery has a more or less fixed voltage and provides variable amounts of current; but at a given insolation level the solar cell provides a more or less fixed current over a wide range of voltage.

The maximum voltage of the cell, its *open-circuit voltage* V_{oc}, is given by the intercept on the voltage axis and lies in the range 0.5 V–0.6 V. It does not depend greatly on the insolation. The close relationship between the diode characteristic of the *p–n* junction and the *I–V* characteristics in sunlight, illustrated in Figure 2.10(a), means that the open-circuit voltage is similar to the forward voltage of about 0.6 V at which a silicon diode starts to conduct heavily.

The maximum current from the cell, its *short-circuit current* I_{sc}, is given by the intercept on the current axis and is proportional to the strength of the sunlight. Other things being equal it is also proportional to the cell's surface area. It represents the full flow of minority carriers generated by the sunlight and successfully 'collected' after crossing the *p–n* junction.

The above parameters are further illustrated by Figure 2.11. The blue curve shows a typical *I–V* characteristic at 1000 W/m^2 insolation, labelled with the short-circuit current, open-circuit voltage, and maximum power point. The red curve shows how power output varies with voltage; the maximum value is $P_{mp} = I_{mp} \times V_{mp}$. Since the current holds up well over most of the voltage range, it follows that the cell's output power is roughly proportional

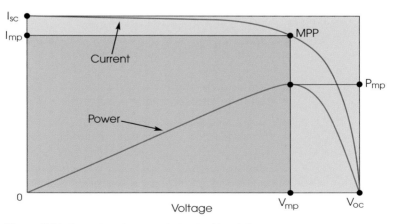

Figure 2.11 Current and power at standard insolation.

to voltage up to the MPP. This emphasises once again the importance of operating the cell close to the MPP if its power output potential is to be realised.

A widely used measure of performance that reflects the overall quality of the cell is its *fill factor* (*FF*) given by:

$$FF = (I_{mp}V_{mp})/(I_{sc}V_{oc}) \qquad (2.4)$$

An 'ideal' cell in which the current held right up to the short-circuit value, then reduced suddenly to zero at the MPP, would have a fill factor of unity. Needless to say, practical cells do not achieve this; the *I–V* characteristic in the figure has a fill factor of about 70%. Equation (2.4) shows that graphically it is equal to the ratio between the areas of the small and large shaded rectangles in the figure.

So far we have not considered the effects of temperature on cell performance, but actually they are quite important, especially in the case of crystalline silicon. Many people imagine that solar cells are more efficient if operated at elevated temperatures, perhaps thinking of the type of solar-thermal panel used for water heating. But solar photovoltaic cells like to be kept cool – they do very well in strong winter sunshine in the Swiss Alps! In hot climates cell temperatures can reach 70 °C or more and system designers often go to considerable lengths to ensure adequate ventilation of PV modules to assist cooling.

The main effect of temperature on a cell's *I–V* characteristic is a reduction in open-circuit voltage, illustrated by Figure 2.12. We have repeated the 1000 W/m^2 curve for the 2 W$_p$ cell already shown in Figure 2.10(b) for the standard temperature of 25 °C, and added two further curves for 0 and 50 °C. The open-circuit voltage changes by about 0.1 V between these extremes, corresponding to 0.33% per °C. Note that the *temperature coefficient* is negative; in other words the voltage decreases as the temperature rises. There is a much smaller effect on the short-circuit current. Generally the cell loses power at elevated temperatures, a more serious effect with crystalline silicon than most other types of solar cell.

You have probably noticed one major omission from this discussion – an explanation of efficiency. At the start of this chapter we noted that commercial crystalline silicon modules have typical efficiencies in the range 11–16%, but we have not so far explained the reasons for this apparently rather disappointing performance. Returning for a moment to Figure 2.10(b) it is not clear from our discussion why this cell, which probably receives up to about 14 W$_p$ of incident solar energy, only manages to convert 2 W$_p$ into electrical output. Where does the rest go, and why can't

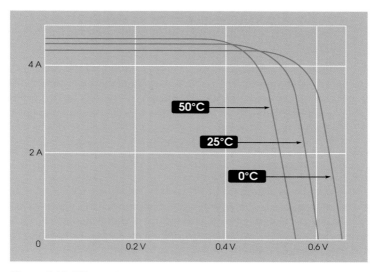

Figure 2.12 Effects of temperature on the *I–V* characteristic.

the efficiency be dramatically improved by better design? This raises some fundamental issues which we tackle in the next section.

2.2.3.3   Sunlight, silicon, and quantum mechanics

It may seem a little surprising to find 'quantum mechanics' mentioned in an introductory book on photovoltaics – and possibly unnerving in view of a quotation by Richard Feynman (1918–1988), latterly a professor at the California Institute of Technology, who received a Nobel Prize in Physics in 1965 for his work on quantum mechanics and famously declared: 'I think I can safely say that nobody understands quantum mechanics'.

So it is clear we must tread lightly, leaving the great body of 20th century quantum theory undisturbed. Yet not entirely, for it contains precious nuggets relating to the nature of sunlight and imposes fundamental limits on the efficiency of solar cells.[1,2]

Back in Section 1.4 we noted that certain eminent physicists, from Isaac Newton in the 17th century to Albert Einstein in the 20th, viewed light as a stream of minute particles carrying discrete packets of energy. And in Section 2.2.3.1 we stated – without explanation – that a light quantum or photon needs a certain minimum energy, known as the *bandgap*, if it is to

43

have any chance of creating an electron–hole pair in a silicon crystal lattice. It is now time to bring these ideas together with the help of a little quantum theory.

The human eye is sensitive to visible light – all the colours of the rainbow from violet to red. The corresponding range of wavelengths is about 0.4 to 0.8 μm. The complete solar spectrum, previously shown in Figure 1.4, also contains significant energy at ultraviolet (UV) and especially infrared (IR) wavelengths. A key concept of quantum theory is that the energy content of a photon is related to wavelength by a surprisingly simple equation:

$$E = hc/\lambda \qquad\qquad (2.5)$$

Where E is the photon energy, h is Planck's constant, c is the velocity of light, and λ is the wavelength. This means that the packet of energy or quantum is about twice as large for a violet photon as for a red photon. And as Einstein proposed in 1905, quanta can only be generated or absorbed as complete units.

A second key point is that solar cells based on semiconductors are essentially quantum devices. An individual solar photon can only generate an electron–hole pair if its quantum of energy exceeds the bandgap of the semiconductor material, also known as its *forbidden energy gap*. This is illustrated by Figure 2.13.

You may recall that the creation of an electron–hole pair involves jolting a valence electron to produce a broken bond in the crystal lattice. The electron moves from the *valence band* to the *conduction band*, leaving

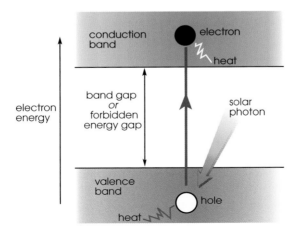

Figure 2.13 Quantum effects in solar cells.

behind an equal, but oppositely charged hole. However the energy levels of an electron in the two bands are separated by a discrete energy gap. Moving from one band to another requires a 'quantum leap' – it is all or nothing, and intermediate levels are forbidden. Long-wavelength infrared and red photons do not generally have the necessary amount of energy. Conversely most photons towards the violet end of the spectrum have more than enough and the excess must be dissipated as heat. These fundamental considerations, taken in conjunction with the Sun's spectral distribution, reduce the theoretical maximum efficiency of a silicon solar cell at an insolation of $1000\,W/m^2$ to about 45%. The figure does not take account of various other loss mechanisms and practical design considerations, some of which were illustrated by Figure 2.8. So it is not hard to appreciate why cells made in research laboratories do well to reach 30% and why current commercial, mass-produced, cells achieve less than 20%.

We can now appreciate why the size of the bandgap is a very important influence on solar cell efficiency. If the bandgap is too large many photons possess insufficient energy to create electron–hole pairs. But if it is too small, many have a lot of excess energy that must be dissipated as heat. It is found that efficient harvesting of the Sun's energy requires bandgaps in the range 1.0–1.6 electron volts (eV). Silicon's bandgap of 1.1 eV is fairly good in this respect. Certain other semiconductor materials have bandgaps closer to the middle of the range, and we will discuss them later.

Unfortunately not all photons with the necessary energy are readily absorbed. Most solar cell materials, the *direct-bandgap* semiconductors, act as good light absorbers within layers just a few micrometres thick. But crystalline silicon, an *indirect-bandgap* material, is not so effective. It absorbs high-energy blue photons quite easily, close to the cell's top surface, but low-energy red photons generally travel much further before absorption and may exit the cell altogether. The basic problem is that successful generation of conduction electrons in silicon requires additional quantum lattice vibrations that complicate the process, so that layers less than about 1 mm thick are not good light absorbers. Special *light-trapping* techniques may be used to increase the pathlength of light inside the cell and give a better chance of electron–hole generation. These are described in the next section.

To summarise, it would be helpful if every photon entering a solar cell produced an electron–hole pair and contributed to power generation, in other words if the *quantum efficiency* was 100%. But quantum theory tells us this is impossible. Photons are all-or-nothing packets of energy that can only be used in their entirety. Some are too feeble in their energy content while others are unnecessarily strong, placing fundamental limits on solar cell efficiency. Disappointing though this may seem, we should always remember that sunlight is 'free' energy, to be used or not as we wish. Photons are not wasted if untapped – at least not in the sense of an old-fashioned power station burning fossil fuel that effectively discards around 60% of its precious fuel as waste heat.

2.2.3.4 Refining the design

Solar cell designers are constantly striving to improve conversion efficiencies, and have used their ingenuity over many years to refine crystalline silicon cells beyond the basic scheme already illustrated in Figure 2.8. Some of the constraints on efficiency are caused by fundamentals of light and quantum theory, others by the properties of semiconductor materials or the problems of practical design.

One important point should be made at the outset. Researchers use various sophisticated techniques to achieve 'record' efficiencies and can select their best cells for independent testing and accreditation. But PV companies engaged in large-scale production have an additional set of priorities: simple, reliable, and rapid manufacturing processes; high yield coupled with minimal use of expensive materials, all aimed at lower costs. Manufacturers are certainly interested in the commercial advantages of high cell efficiency and over the years have incorporated many design advances coming out of research laboratories, but cost must always be a big consideration and there are often significant time lags.

Figure 2.14 summarises the main factors determining the efficiency of a typical, commercial, crystalline silicon solar cell operated at or near its maximum power point. On the left the incident solar power is denoted by 100%. Successive losses, shaded in blue, reduce the available power to around 15–20% at the cell's output terminals – its rated efficiency value. We will now discuss each loss category in turn.

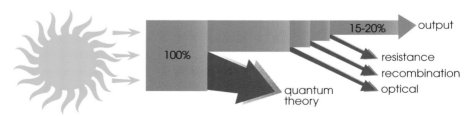

Figure 2.14 Solar cell losses.

Quantum theory

We emphasised the fundamental limitations imposed by quantum theory[4] in the previous section. They represent the biggest loss of efficiency in a solar cell based on a single *p–n* junction. One way of reducing the problem is to stack together two or more junctions with different bandgaps, creating a *tandem cell*. A well-known example, which has been exploited commercially for many years, is based upon amorphous rather than crystalline silicon and we shall mention this again in Section 2.3.

Optical losses

Optical losses affect the incoming sunlight, preventing absorption by the semiconductor material and production of electron–hole pairs. The small section of solar cell shown in Figure 2.15 illustrates three main categories of optical loss: blocking of the light by the top contact (1); reflection from the top surface (2); and reflection from the back contact without subsequent absorption (3).

Shadowing by the top contact can obviously be minimised by making the total contact area as small as possible. This area comprises not only the metallic contact fingers shown in the figure (and previously in Figure 2.8) but also wider strips known as *busbars* that join many fingers together and conduct current away from the cell. Clearly a well-spaced grid of very fine

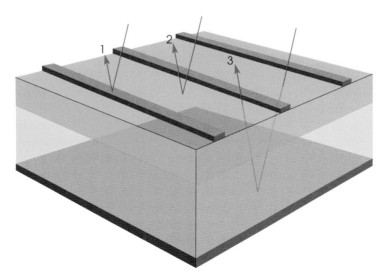

Figure 2.15 Optical losses.

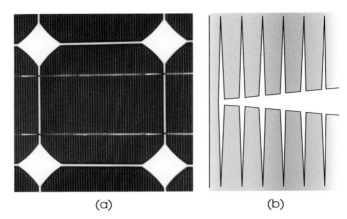

Figure 2.16 Contact fingers and busbars.

fingers and narrow busbars helps reduce optical loss, but the disadvantage is increased electrical resistance. As always, practical design involves compromise.

The photo in Figure 2.16 shows the top surface of a monocrystalline silicon cell, surrounded by its neighbours in a PV module. This example has a very simple grid geometry, consisting of 49 fine vertical fingers and two horizontal busbars, giving a shadowing loss of about 11%. The fingers have constant width; a more efficient design would taper them to account for the increasing current each carries as it nears a busbar. The busbars are slightly tapered towards the low-current end; it would be better to taper them along their length as they pick up current from more and more fingers. Ideally the cross-sections of fingers and busbars should be roughly proportional, at each point, to the current carried. To illustrate this a small section of a more efficient finger-busbar design is shown in part (b) of the figure.

The *metallisation pattern* of fingers and busbars, as well as having its own inherent resistance to current flow, introduces contact resistance at the semiconductor interface. This may be reduced by heavy doping of the top layer of semiconductor material, at the risk of forming a significant dead region at the surface that reduces the collection efficiency of blue photons.

Conventional top contacts are made from very thin metallic strips formed using a screen-printing process. A metallic paste is squeezed through a mask, or screen, depositing the desired contact pattern which is then fired. The shading loss, typically between 8 and 12%, represents a significant

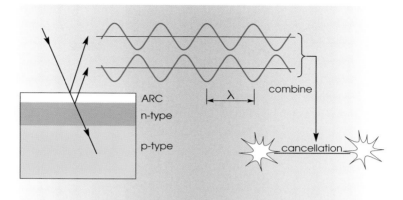

Figure 2.17 An antireflection coating reduces reflection from the top surface by cancellation.

drain on cell efficiency. A major design improvement, pioneered in the 1990s at the University of New South Wales,[3] uses laser-formed grooves to define a metallisation pattern with narrower but deeper fingers just below the cell's surface. Such *buried contact solar cells* offer valuable gains in efficiency compared with normal screen-printed designs.

The second category of optical loss illustrated in Figure 2.15 is reflection from the cell's top surface. Two main design refinements are commonly employed. The first is to apply a transparent dielectric *antireflection coating* (*ARC*) to the top surface, illustrated by Figure 2.17. If the coating is made a quarter-wavelength thick, the light wave reflected from the ARC/silicon interface is 180° out of phase with that reflected from the top surface and when the two combine the resulting interference effects produce cancellation. This condition is met when:

$$d = \lambda/4n \qquad (2.6)$$

where d is the thickness and n the refractive index of the coating material, and λ is the wavelength (interestingly, we are temporarily considering light as a wave rather than a stream of particles, a good example of the dual nature of light first mentioned in Section 1.4). Clearly, exact cancellation can only occur at one value of λ, normally chosen to coincide with the peak photon flux at about 0.65 μm. The antireflection performance falls off to either side of this value. For optimum performance the refractive index of

Figure 2.18 Texturisation by raised pyramids.

the ARC material should be intermediate between that of the materials on either side, usually silicon and either air or glass.

The second design refinement involves *texturising* the top surface so that light is reflected in a fairly random fashion and has a better chance of entering the cell. Almost any roughening is helpful, but the crystalline structure of silicon offers a special opportunity because careful surface etching can be used to create a pattern of minute raised pyramids, illustrated in Figure 2.18. Light reflected from the inclined pyramidal faces is quite likely to strike adjacent pyramids and enter the cell.

The third type of optical loss is reflection of light from the back of the cell, without subsequent absorption. This may be reduced by an uneven back surface that reflects the light in random directions, trapping some of it in the cell by total internal reflection. The technique is referred to as *light trapping*[3] and is very important in crystalline silicon cells because silicon is a relatively poor light absorber, especially of longer-wavelength (red) light. It is illustrated in Figure 2.19.

It is difficult to put precise figures on the efficiency losses caused by these various optical effects. However a cell that includes carefully designed metallisation, ARC, texturisation, and light trapping can give major improvements compared with the basic structure first illustrated in Figure 2.8.

Recombination losses

The undesirable process known as recombination has already been discussed in Section 2.2.3.1. It occurs when light-generated electrons and holes, instead of being swept across the *p–n* junction and collected, meet up and are annihilated. The wastage of charge carriers adversely affects both the voltage and current output from the cell, reducing its efficiency.

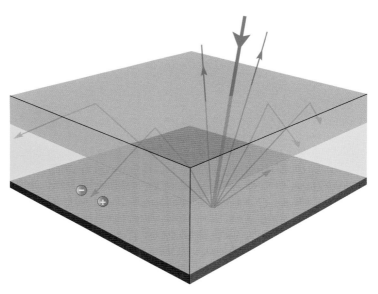

Figure 2.19 Light trapping helps keep incoming light within the cell by total internal reflection.

Some recombination takes place as electrons and holes wander around in the body of the cell (*bulk recombination*), but most occurs at impurities or defects in the crystal structure near the cell's surfaces, edges, and metal contacts, as illustrated in Figure 2.20. The basic reason is that such sites allow extra energy levels within the otherwise forbidden energy gap (see Figure 2.13). Electrons are now able to recombine with holes by giving up energy in stages, relaxing to intermediate energy levels before finally falling back to the valence band. In effect they are provided with stepping stones to facilitate the quantum leaps necessary for recombination.

What can be done to reduce recombination? Three important techniques may be briefly mentioned here. The first involves processing the cell to create a *back surface field* (*BSF*). Although the details are subtle,[4] the tendency of red photons to recombine at the back of the cell may be reduced by including a heavily doped aluminium region which also acts as the back contact. Next, it is possible to reduce recombination at the external surfaces by chemical treatment with a thin layer of *passivating oxide*. And finally, regions adjacent to the top contacts may be heavily doped to create 'minority carrier mirrors' that dissuade holes in the *n*-type top layer from approaching the contacts and recombining with precious free electrons.

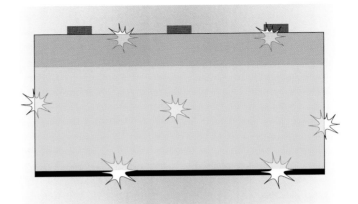

Figure 2.20 Typical recombination sites. The central one represents bulk recombination, the others occur close to surfaces, edges, and contacts.

Resistance losses

The final efficiency loss shown in Figure 2.14 is due to electrical resistance. We previously noted that a solar cell is best thought of as a current generator. As with other current generators, it is desirable to minimise resistance in series with the output terminals and maximise any shunt resistance that appears in parallel with the current source. Figure 2.21 shows two equivalent circuits similar to that previously used for a solar cell (Figure 2.9) but modified to include a series resistance R_1 in part (a) and a shunt resistance R_2 in part (b). Ideally, R_1 would be zero and R_2 infinite, but needless to say, we cannot expect these values in practice.

The physical interpretation of R_1 is straightforward. It represents the resistance to current flow offered by the busbars, fingers, contacts and the cell's bulk semiconductor material. A well-designed cell keeps R_1 as small as possible. R_2 is more obscure, relating to the nonideal nature of the p–n junction and impurities near the cell's edges that tend to provide a short-circuit path around the junction. In practical designs both resistors cause losses, but it is simpler to appreciate their effects if we treat them separately.

The black I–V characteristic in part (a) is for $R_1 = 0$, the ideal case, which we refer to as the reference cell. The red characteristic is for a cell with a finite value of R_1. Let us first consider the open-circuit condition, $I = 0$. In this case there is no current through R_1 and no voltage drop across it, so the open-circuit voltage V_{OC} must be the same as for the reference cell. We

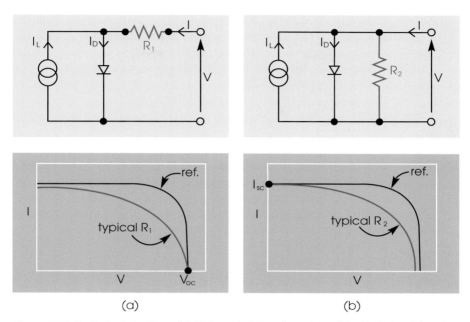

Figure 2.21 Equivalent circuits and *I–V* characteristics of a solar cell that includes: (a) series resistance; (b) shunt resistance.

conclude that series resistance due to a cell's busbars, fingers, contacts and semiconductor material has no effect on the open-circuit voltage. However, full circuit analysis shows that it causes a small reduction in short-circuit current and a considerable loss of fill factor, as indicated.

Part (b) of the figure shows the effects of shunt resistance and it is helpful to consider the short-circuit condition, $V = 0$. In this case there is no voltage across R_2 and no current through it, so the short-circuit current I_{SC} must be the same as for the reference cell. We conclude that finite shunt resistance due to imperfections in and around the cell's *p–n* junction has no effect on the short-circuit current. However, it has a minor effect on the open-circuit voltage and a considerable one on the fill factor. To conclude, a practical cell with both series and shunt resistance losses is expected to suffer small reductions in both V_{OC} and I_{SC}; but the most serious effect is generally degradation of fill factor.

We have now covered the main categories of efficiency loss in crystalline silicon solar cells. The techniques for counteracting them have been conceived and enhanced over many years in R&D laboratories around the world, leading to continuous improvements in cell and module efficiencies. Of course, the degree to which they are employed in a commercial product depends upon the manufacturer's expertise and judgement; the number and complexity of processing steps have a big impact on cost and there is inevitably a trade-off between cost and performance.

2.2.4 Multicrystalline silicon

In most respects multicrystalline silicon, also referred to as *polycrystalline silicon* or more simply as *poly-Si,* produces solar cells that are very similar to their monocrystalline cousins. The theoretical background is shared, even though the initial stage of manufacture is different. As first mentioned in Section 2.1, multicrystalline cells also start life as pure molten silicon, but the material is cast in substantial blocks, cut into smaller bricks, and finally made into thin wafers. The casting process produces a multi-grain crystal structure that is less ideal than monocrystalline material and gives cell and module efficiencies typically 1% (absolute) lower, but this dis-advantage is offset by lower wafer costs. And since the cells are cut square or rectangular, rather than 'pseudo-square' as with monocrystalline cells, they can be packed closely in modules. They have a scaly, shimmering appearance. The cable-car station previously illustrated in Figure 2.2 shows that the modules tend to have a distinctly blue appearance due to their antireflection coatings, a property often appreciated by architects.

As the molten silicon cools, crystallisation occurs simultaneously at many points, producing crystal grains with random sizes, shapes and orientations. After cutting into thin wafers, the material has the surface appearance of Figure 2.22(a). Within each grain the crystal structure is highly regular, but the many grain boundaries represent imperfections and provide unwelcome sites for electron–hole recombination. The problem is reduced if grains are at least a few millimetres across and extend from front to back of the wafer. As part (b) of the figure shows, a multicrystalline module tends to present a uniform, shimmering, appearance without the gaps between cells associ-ated with the 'pseudo-square' shape of monocrystalline cells.

On the whole there is little to choose between the performance of mono-crystalline and multicrystalline PV modules. From a user's point of view efficiency and cost differences may not be decisive and the choice often comes down to appearance, availability, and the manufacturer's reputation and guarantee.

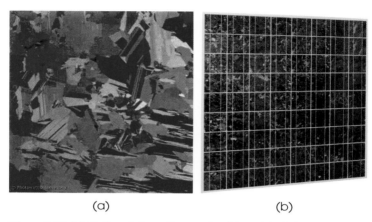

(a)   (b)

Figure 2.22 (a) Multicrystalline silicon wafer; (b) module (EPIA/Photowatt).

2.3 Amorphous and thin-film silicon

Amorphous silicon (a-Si) was the first *thin-film* technology used in PV. Small a-Si cells in consumer products such as watches and calculators have introduced solar cells to millions of people since the 1980s. The tiny amounts of power required by such products make the comparatively low efficiency of their cells unimportant, and in any case they are rarely used out of doors in strong sunlight! Ease of manufacture and low cost are their strong points. What is not so generally realised is that a-Si technology has been developed in recent years and scaled up for higher-power applications. Although it only accounts for a few percent of world production, it is no longer confined to consumer products. A good example is building façades; a-Si modules can serve as attractive cladding and may well be competitive with other types of PV module. PV cladding is not necessarily more expensive than traditional high-quality materials and may be chosen for its aesthetic appeal, or as an environmental statement. If the façade also generates electricity, so much the better. Efficiency is not the only criterion.

In any case the question of efficiency needs further discussion. We noted at the start of this chapter that a-Si module efficiencies typically fall in the range 6–8%, about half that of crystalline silicon. But efficiencies quoted by PV manufacturers invariably relate to standard insolation ($1000 \, \text{W/m}^2$, $25 \, °\text{C}$) and tell only part of the story. While crystalline silicon modules are

55

Figure 2.23 Amorphous silicon PV modules on a building façade (EPIA/Schott Solar).

impressive in strong sunlight, their performance in weak or diffuse light is often inferior to thin-film products and is more adversely affected by high temperatures. In recent years there have been many reports of thin-film modules outperforming crystalline silicon in terms of annual energy yield, especially in climates with significant cloud cover and plenty of diffuse light.

Amorphous silicon is also a far better light absorber than crystalline silicon, so extremely thin layers of semiconductor may be used – of the order 1 μm. Like other thin-film technologies it offers further advantages:

- relatively simple fabrication at low temperatures using inexpensive substrates and continuous 'production line' methods;
- integrated, monolithic, design obviating the need to cut and mount individual wafers;
- potential for manufacturing flexible, lightweight products.

The word *amorphous*, derived from ancient Greek, means 'without form or shape'. Amorphous silicon (a-Si), which may be deposited as a thin film on a variety of substrates, does not exhibit a regular lattice structure. The distances and angles between the silicon atoms are randomly distributed, giving rise to incomplete bonds and a high concentration of defects. The result is a high density of allowed energy states within the nominal energy gap, in stark contrast to crystalline silicon (see Figure 2.13). In effect, the extra energy states act as stepping stones, allowing conduction electrons to relax back into the valence band and recombine. There is also a problem of low charge-carrier mobility within the semiconductor material (referred to as poor *carrier transport*). Fortunately, early research into a-Si solar cells suggested two effective ways of countering these difficulties.

First, it was discovered that introducing hydrogen into amorphous silicon could passivate incomplete bonds, also known as *dangling bonds*, greatly reducing the number of excess energy states within the bandgap. The modified material is referred to as a-Si(H) to denote its hydrogen content and is illustrated by Figure 2.24. This shows the irregular arrangement of silicon atoms, a dangling bond (DB), and a dangling bond that has been passivated by a hydrogen atom (H). Using this approach it is possible to make effective *n*-type and *p*-type material by doping with phosphorus or boron, resulting in a direct bandgap semiconductor with an energy gap of about 1.75 eV.

The second problem, poor carrier transport, is reduced by introducing an intrinsic layer (which, in practice, is usually slightly *n*-type) into the *p–n* junction giving the *p–i–n* structure shown in Figure 2.25. This *i*-layer greatly increases the width of the depletion region and the associated

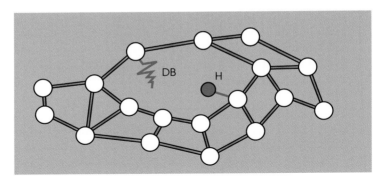

Figure 2.24 Irregular structure and bonding in a-Si(H).

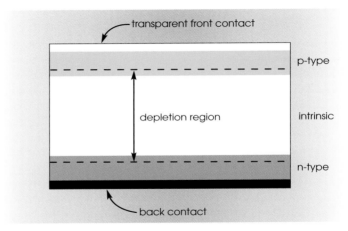

Figure 2.25 The basic structure of a single-junction a-Si(H) solar cell.

electric field that sweeps minority carriers across the junction. Assuming the *i*-layer is in fact lightly doped *n*-type, the highest electric field occurs at the *p–i* interface and it is therefore best to design the cell so that light enters through a transparent front contact into a very thin, heavily doped, *p*-type layer. This ensures that most charge carriers are created near the top of the cell and successfully collected.

Unfortunately, the introduction of an *i*-layer has its drawbacks. During initial exposure to strong sunlight, absorption by the *i*-layer creates additional defects that aid recombination and reduce cell efficiency. The phenomenon, known as the *Staebler–Wronski effect*, depends on the total number of photons absorbed and therefore on the intensity and duration of the light and the thickness of the *i*-layer. Building up over a timescale of months, it results in final or 'stabilised' efficiencies significantly lower than the initial values. In the past this has given single-junction a-Si(H) cells a rather doubtful reputation.

But most PV clouds have a silver lining. In the case of Staebler–Wronski, the initial loss of efficiency can be largely overcome using multi-junction or stacked cell structures in which light absorption is shared between two or more much thinner *i*-layers. Furthermore, by stacking cells with different bandgaps it is possible to capture a bigger percentage of solar photons and achieve relatively good levels of efficiency and stability, especially in weak or diffuse sunlight.

The basic scheme for one type of triple-junction cell is shown in Figure 2.26. It depends on the ability of a-Si to form good alloys with germanium, producing semiconductor material with smaller bandgaps. The top a-Si 'blue cell' is effective at capturing high-energy blue photons with its bandgap of about 1.75 eV. Next comes the 'green cell', based on amorphous silicon–germanium alloy containing about 15% germanium with a bandgap of around 1.6 eV. And finally the bottom 'red cell', designed to capture low-energy red and infrared photons, uses an alloy with about 50% germanium giving a bandgap of around 1.4 eV. Photons that are not absorbed on first pass through the cells are returned by the back reflector which may be texturised to encourage light-trapping.

The supporting substrate does not have to be flexible, but flexibility offers exciting possibilities during production and also for the user. The production process can be continuous 'roll-to-roll', the various layers being deposited on an extremely long thin sheet of stainless steel or plastic as it travels

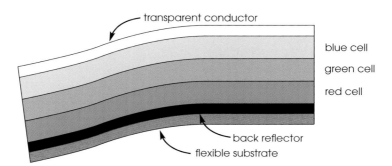

Figure 2.26 A triple-junction amorphous silicon solar cell.

Figure 2.27 Roll-to-roll manufacture of a-Si solar cells (IEA-PVPS).

between rollers in the manner of a magic carpet. This was the dream of solar cell pioneers back in the 1950s! Sheet thickness is typically a small fraction of a millimetre, with sheet lengths up to an amazing several kilometres. Individual solar cells are automatically scribed and interconnected as a monolithic circuit. From the user's point of view, flexibility tends to go hand in hand with lightness and allows easy mounting on curved or awkward surfaces.

The lack of a crystal structure in amorphous silicon ultimately prevents it from matching the efficiency of crystalline silicon, at least in strong sunlight. However recent years have seen much R&D effort directed towards a new microcrystalline form of silicon that, like other thin film materials, can be deposited in extremely thin layers of about $1\,\mu m$ onto various substrates including glass. Crystalline silicon's comparatively poor light absorption means that success depends upon highly effective light-trapping to keep incident light within the film. The hope is that microcrystalline silicon will rival wafer technology for ruggedness and electrical stability, while at the same time using minimal amounts of cheap and plentiful raw materials, improving efficiency above amorphous products, and greatly reducing costs. The thin-film silicon story that started more than a generation ago is far from over!

2.4 Other cells and materials

Silicon and germanium may be the best-known semiconductors, but they are certainly not the only ones. Many compounds incorporating rather unfamiliar chemical elements also display electrical properties midway between insulators and conductors. Some readily absorb solar photons to produce electron–hole pairs, may be doped to make *n*-type or *p*-type material and deposited as thin layers on a variety of substrates. In other words they are candidates for 'second-generation' thin-film cells that surpass amorphous silicon's efficiency and challenge crystalline silicon's usage of materials and production costs. Of the various possibilities, two materials with exotic names – *copper indium diselenide* (*CIS*) and related compounds, and *cadmium telluride* (*CdTe*) – already have a highly significant presence in the terrestrial PV market and, together with the thin-film microcrystalline silicon mentioned at the end of the previous section, look set to lead PV decisively into a new era.

Not that crystalline silicon cells will be easily displaced. Global production continues apace. Gigawatts of wafer-based modules are already installed and will generate electricity for many years to come, catching the public eye as ambassadors for PV around the world. However it is generally accepted that thin-film technology is the way forward and offers the best chance of achieving grid parity with conventional electricity generation on a large scale. It looks as if 30% or more of global annual PV production may be thin-film within the next decade.

Although we focus initially on CIS and CdTe, there are many other types of PV cell – for example, those used in spacecraft and in solar concentration systems. Other more exotic technologies are under investigation and some are in commercial production. We will meet a few examples later in this chapter.

2.4.1 Copper indium diselenide (CIS)

To be successful, inorganic crystalline solar cell materials need two essential properties. They must be good light absorbers, turning solar photons into electron–hole pairs; and they must include an efficient *p–n* junction to sweep light-generated minority carriers across the junction and force current through an external circuit.

Many years ago it was discovered that the compound semiconductor copper indium diselenide (CIS) offers excellent light absorption in small-grained layers a micrometre or two thick. Although the electronic and chemical properties of CIS and related compounds are subtle and complex,[5] a few key points can be made here. First, and unlike silicon, CIS cannot be doped to form an efficient *p–n* junction on its own (it cannot form a *homojunction*); but it can be interfaced with another semiconductor, *cadmium sulphide* (*CdS*), to produce an effective *heterojunction*. CIS and CdS are well matched and do not suffer excessive recombination at the interface. Since CdS can only be successfully doped as *n*-type material, the CIS must be doped *p*-type. It is rather difficult to make good metallic contact with CIS; gold is effective, but expensive, so molybdenum is normally used as a back contact.

There is a further important twist to the story. In the 1970s it was discovered that the rather low bandgap of CIS (about 1.1 eV) may be increased by substituting some gallium in place of indium. By varying the gallium content a range of bandgaps relevant to PV cells can be obtained, from about 1.1 eV (no gallium) up to 1.7 eV. In addition, the low open-circuit voltage of CIS is raised towards 0.5 V, comparable to crystalline silicon, meaning that fewer cells need be interconnected to achieve useful module voltages. The modified material, *copper indium/gallium diselenide* (*CIGS*), has achieved many cell efficiency records (it is worth noting that the initials CIS and CIGS tend to be used interchangeably, which can lead to a certain amount of confusion). CIGS passed the 20% efficiency milestone for laboratory cells in 2008. At that stage commercial module efficiencies were already attaining 10–12%, comfortably beating amorphous silicon and within aiming distance of crystalline silicon.

The basic scheme of a typical CIGS cell is shown in Figure 2.28. Light enters the cell via a transparent conducting layer acting as the top contact. Next comes an extremely thin layer of CdS that forms a *p–n* heterojunction with the thicker (but still very thin!) CIGS absorber. A metallic layer, normally molybdenum, provides the back contact and completes the electronic design. The doping of the *p*-type absorber is often graded, being lightest near the junction. This extends the depletion region and its associated electric field well into the absorber where most charge carriers are generated and helps sweep them across the junction. Not shown in the

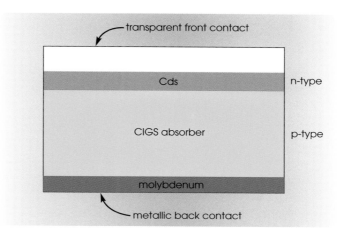

Figure 2.28 The basic scheme of a CIGS solar cell.

figure is the necessary supporting substrate, which may be rigid or flexible and made of glass, metal, or plastic.

As our attention moves from silicon cells with their superabundance of cheap raw material to thin-film cells based on unfamiliar elements, it is time to question cost and availability of supplies. Cost is not generally seen as a problem, given the tiny amounts of material used in thin-film cells compared with silicon wafers; indeed one of thin-film technology's main promises is to make PV ever more affordable. But the situation could change if production levels continue to increase dramatically. The indium used in CIS and CIGS cells is a case in point: indium is a comparatively rare element of the Earth's crust, in demand for electronic products other than solar cells. Availability may become a problem. One advantage of partially substituting gallium into CIGS is a decreased demand for indium, but it is hard to predict how this situation will play out in the medium to long term.

Toxicity is another important issue. Silicon is benign; but cadmium, a heavy metal with a bad reputation as a cumulative poison, is certainly not. So CdS, a compound of cadmium and sulphur, is seen by many as a rather unfortunate component of the p–n junctions used in CIS and CIGS cells. Indeed, considerable efforts have been made to reduce or eliminate cadmium, for example by incorporating atomic sulphur in so-called CIGSS cells, allowing products to be labelled 'cadmium-free'. The PV industry is

Figure 2.29 An array of CIS solar modules in Austria (EPIA/Shell Solar).

extremely aware of its environmental credentials and of public concerns over pollution. It has every reason to minimise risks during manufacture, use, and eventual recycling or disposal.

As thin-film solar cells contribute more and more to 'second-generation' PV technology and challenge the pole position occupied for so long by crystalline silicon, we will become used to seeing CIS and CIGS modules with the smooth, dark grey/black appearance often favoured by architects. There is also intensive development of semitransparent modules that act as windows allowing a portion of light to enter a building while at the same time generating electricity. The possibilities for exciting and innovative PV products are enormously increased by thin-film techniques.

2.4.2 Cadmium telluride (CdTe)

Cadmium telluride (CdTe) is another important semiconductor material for thin-film solar cells, its direct bandgap of 1.45 eV being close to optimum for capturing the Sun's spectrum using a single-junction device. Initially there was considerable concern among environmental groups about the commercialisation of CdTe cells and modules because of cadmium's reputation as a cumulative poison. However, these fears seem to have been largely allayed. CdTe has not got the toxicity of its individual constituents Cd and Te, although there is some lingering concern over fire risk. Cadmium is commonly obtained as a byproduct of zinc mining and smelting, so removing it from the environment for use in solar cells may be seen as an environmental benefit, provided great care is taken over eventual recycling. Cadmium and tellurium are more abundant elements than the indium used in CIS/CIGS products, so availability is not so big an issue – at least not at present production levels. However the market is growing strongly. CdTe modules accounted for over 6% of world production in 2008, more than any other thin-film technology, and they are finding large scale application in PV power plants. Comparatively simple production processes mean that CdTe modules are currently about the cheapest on the market in terms of price per peak watt. Furthermore their conversion efficiencies of around 11% look set to advance towards 15% in the next few years.

The rationale behind a thin-film CdTe solar cell results in a scheme very similar to that for CIS and CIGS. The essential layers in the thin-film 'sandwich' are a transparent top contact, a CdS/CdTe *p–n* heterojunction and absorber, and a metallic back contact, as shown in Figure 2.30. Also required is a suitable supporting substrate of glass, metal, or plastic, that

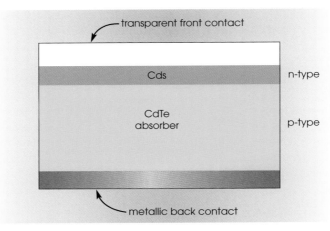

Figure 2.30 A CdTe solar cell.

determines whether cells are rigid or flexible. Bear in mind that although the figure represents the cell as rather thick and narrow, it is actually manufactured as part of an extremely thin sheet.

As worldwide thin-film production grows, and eventually overtakes crystalline silicon, it seems that cadmium telluride will continue its important contribution. An installation that nicely illustrates the possibilities for 'farming sunshine' alongside conventional crops is shown in Figure 2.31. Further up the power scale, a $10\,MW_p$ power plant containing 167 000 CdTe modules with an efficiency of about 11% was recently commissioned near Boulder City, Nevada. Towards the end of 2009 plans were announced for a huge plant in Mongolia that will eventually comprise many millions of modules, a project almost unimaginable a decade ago.

2.4.3 Specialised and innovative cells

2.4.3.1 Gallium arsenide (GaAs)

Gallium is one of the elements in Group III of the Periodic Table; arsenic is in Group V. So gallium arsenide (GaAs) is often referred to as a *Group III–V* semiconductor. GaAs and associated compounds have two claims on our attention as specialised, but important, PV materials: for making solar cells used in spacecraft; and for their use in terrestrial concentrator systems that focus sunlight using mirrors or lenses.

Figure 2.31 Farming the Sun: part of an 810 kW$_p$ CdTe power plant in rural Germany (First Solar/Phoenix Solar).

In the early years of space exploration silicon solar cells were the main source of electricity for spacecraft, reaching efficiencies of about 15% by 1970. Since then GaAs has made a big impact, for two main reasons. First, it is less susceptible than silicon to damage by radiation in space, a key consideration on long missions where the performance and reliability of electricity supply is paramount. Second, its direct bandgap of 1.42 eV (compared with 1.1 eV for silicon) allows a greater percentage of the solar spectrum to be harvested, giving better conversion efficiencies. Since the 1980s solar cell designers have learned how to deposit thin films on crystalline germanium wafers, producing lightweight multijunction devices of even higher efficiency. Triple-junction modules have gained a high reputation for their reliability and light weight. And although the material and processing costs of GaAs cells are high, this is hardly a major consideration for vastly expensive space projects.

A typical scheme for triple-junction GaAs cells is shown in Figure 2.33. Like the triple-junction amorphous silicon cell described earlier, it is a "sandwich" of three stacked cells with different bandgaps designed to capture different portions of the Sun's spectral energy. In this case the relevant spectrum corresponds to *air mass zero* (*AM0*), received by solar cells outside the Earth's atmosphere (refer back to Figure 1.6). Each cell

67

Figure 2.32 Preparing for launch: a large Space PV array (Boeing/Spectrolab).

includes *n*-type and *p*-type crystalline layers. The top cell, with a bandgap of about 1.9 eV obtained using the alloy *gallium indium phosphide (GaInP)*, is very effective at absorbing high-energy UV/blue photons. The GaAs cell in the middle has a bandgap of 1.42 eV; and the bottom cell, based on germanium that also provides the supporting substrate, has a bandgap of 0.7 eV to absorb infrared photons.

Although such triple-junction devices come in the general category of 'gallium arsenide', we see that they actually use carefully controlled proportions of several III–V elements plus Group IV germanium to achieve bandgap control. These highly specialised solar cells are built up monolithically, many layers being grown one on top of one another with optimal thickness and doping. All this requires expensive materials and very advanced processing. But the technical rewards are high: the best laboratory cells have efficiencies well over 30%, with commercial cells and modules not far behind.

Such impressive efficiencies pose an interesting question. Can gallium arsenide be 'brought down to earth' and make a significant contribution to terrestrial PV generation? Success depends upon effective *concentration* of sunlight using mirrors or lenses, focusing the light onto cells of far smaller

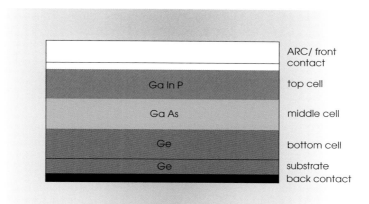

Figure 2.33 A triple-junction cell.

area with correspondingly reduced material and processing costs. For example, increasing the light intensity 500 times ('500 Suns') should allow the cell area to be reduced 500 times for the same power output. Indeed, it is rather better than this because the efficiency of many solar cells improves under concentrated sunlight. Triple-junction GaAs concentrator cells have already passed the 40% landmark in the laboratory, with commercial cells not far behind.

Successful PV concentration systems must aim to reduce cell costs sufficiently to offset the expense of focussing the light and tracking the Sun across the sky on its daily journey. Not surprisingly, there are sceptics; yet PV concentration is being intensively researched and developed, with many systems in commercial production. We shall say more about them in the next chapter.

2.4.3.2 Dye-sensitised cells

Some of the new PV concepts and materials introduced in recent years would have astounded early PV pioneers whose attention was entirely focused on inorganic semiconductors, principally silicon and germanium. We are now moving into an era where artificial organic materials seem certain to play an important role in converting sunlight directly to electricity. They are seen as part of PV's 'Third Generation'. Of many possible approaches, *Dye-sensitised cells* (*DSCs*) are presently in the vanguard of development and commercial application.

The modern DSC broke upon the scene in 1991 when Michael Graetzel and Brian O'Regan at the Federal Polytechnic in Lausanne, Switzerland reported that a 10 μm thin-film of the semiconductor *titanium dioxide* (*TiO₂*) could work as an effective solar cell if coated with an organic dye, immersed in an electrolyte, and provided with electrical contacts.

Most importantly, the TiO₂ was made in the form of a nanoporous 'sponge' of minute particles just tens of nanometres (nm) across, propelling PV into the modern field of nanotechnology. And since titanium dioxide (also known as *titania*) is an inorganic semiconductor whereas the dye and electrolyte are organic, the *Graetzel cell* is sometimes referred to as an organic–inorganic thin film device.

But why *dye-sensitised*? Unlike conventional cells in which the absorption of light and transport of light-generated charges takes place within the same semiconductor, in a DSC these roles are split. The dye acts as light-absorber, generating electrons which it *injects* into the conduction band of the semiconductor. In other words the dye acts as a 'sensitiser' of the TiO₂, which would not be effective on its own because of its large bandgap. Another key aspect of Graetzel cells is their use of new organic dyes able to absorb a wide solar spectrum. And the use of TiO₂ nanoparticles, rather than larger crystals, hugely increases the surface area of the adsorbed dye coating and hence the efficiency of light absorption.

Most people, meeting DSCs for the first time, find their detailed operation very complex – certainly more so than crystalline silicon cells. Although it involves many of the same basic concepts[6] – photon absorption, charge generation and transport, recombination, optical and resistance losses – the electrochemical terminology is unfamiliar and the names of the organic materials can seem unreasonably long! So we restrict ourselves here to a brief summary.

The basic scheme of a DSC is illustrated by Figure 2.34. Light enters the cell via a transparent front contact and is absorbed by the organic dye covering the TiO₂. Excitation electrons are injected into the conduction band of the TiO₂, causing oxidation of the dye. They are efficiently transported through the semiconductor by diffusion and reach the electrical contact. Assuming the cell is connected to an electrical load, the electrons now pass through the external circuit and re-enter the cell via the back contact or *counter-electrode*. Here they provide the negative charges required to restore the dye to its original (unoxidised) state with the help of the intervening electrolyte. The circuit is completed.

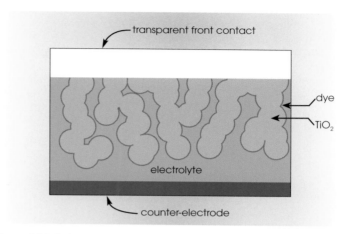

Figure 2.34 A dye-sensitised solar cell.

Unfortunately some recombination does occur, but not in the same manner as in conventional silicon cells. Although electrons are injected by the dye into the conduction band of the semiconductor, holes are not formed in its valence band; so there is no generation of electron–hole pairs – or subsequent annihilation. But electrons can recombine with the oxidised dye. Fortunately, electron injection and transport in the semiconductor is extremely fast compared with the recombination process, so effective charge separation does in fact take place. Overall, the photon-to-electron generation process in a DSC is analogous to photosynthesis in leaves and plants where chlorophyll acts as the sensitiser.

Early Graetzel cells achieved very respectable efficiencies of up to about 10% in standard insolation conditions ($1000\,W/m^2$, $25\,°C$). A great deal of ongoing research has since improved performance, raising efficiency above amorphous silicon and within sight of other thin-film technologies. However efficiency in bright sunshine is probably not the main criterion for DSCs. They work well in low, diffuse light and in high ambient temperatures, indoors and out. Flexible modules can easily be made using plastic substrates. They use nontoxic and plentiful materials (TiO_2 is a widely used chemical – for example in paints and toothpastes) and their relatively simple manufacturing techniques include fast roll-to-roll production. Unusual and exciting possibilities are opening up for building-integrated PV, including roofing products, transparent and semitransparent tinted

Figure 2.35 Innovative and flexible: dye-sensitised solar cells in Australia (Dyesol).

windows, partitions and decorative features. Instead of restricting architects to standard rectangular PV modules, DSC products can be tailor-made to particular sizes, shapes, and aesthetic design criteria. The wide range of applications promises an exciting future.

References

1. T. Markvart (ed.). *Solar Electricity*, 2nd edition, John Wiley & Sons, Ltd: Chichester (2000).
2. P. Wuerfel. *Physics of Solar Cells*, Wiley-VCH: Weinheim (2005).
3. S.R. Wenham *et al. Applied Photovoltaics*, Earthscan: London (2007).
4. A Luque and S. Hegedus (eds). *Handbook of Photovoltaic Science and Engineering*, John Wiley & Sons, Ltd: Chichester (2003).
5. J. Poortmans and V. Arkhipov. *Thin film Solar Cells*, Wiley-VCH: Weinheim (2006).
6. M. Pagliaro *et al. Flexible Solar Cells*, Wiley-VCH: Weinheim (2008).

3 PV modules and arrays

3.1 Introductory

Modules and arrays present PV's face to the World, as well as the Sun, and the technology's reputation depends crucially on their technical performance, reliability, and appearance. They must be designed and manufactured for a long and trouble-free life. The solar cells they contain need careful encapsulation to provide mechanical strength and weatherproofing, and the electrical connections must remain robust and corrosion-free.

Most PV modules are provided with aluminium frames to give extra protection and simplify mounting on a roof or support structure. Modules without frames, known as *laminates*, are sometimes preferred for aesthetic reasons, for example on the façade of a building where reflections from metal frames would be unwelcome. A group of interconnected modules working together in a PV installation is referred to as an *array*. We mentioned PV modules briefly in Section 2.1, noting the module areas required for a given power output using different cell technologies, and discussed cell and module efficiencies. In this chapter we will focus mainly on electrical characteristics and effective mounting to capture the available sunlight. But first, a few words about module sizes and designs.

For a given level of solar cell efficiency, the rated power output of a module is proportional to its surface area. As we noted in Section 2.1, about $7–8\,m^2$ of surface area is required to generate $1\,kW_p$ using crystalline silicon cells, about $16\,m^2$ using amorphous silicon, and intermediate areas for thin-film

Electricity from Sunlight   By Paul A. Lynn
© 2010 John Wiley & Sons, Ltd

Figure 3.1 A large array of PV modules on a rooftop in Switzerland (EPIA/BP Solar).

technologies such as CIGS and CdTe. As more and more PV installations including power plants move into the megawatt range, huge arrays and module numbers are involved. For example the $10\,MW_p$ system mentioned in Section 2.4.2 uses 167 000 CdTe modules with a total area of some $100\,000\,m^2$. The designers of such systems normally specify the largest available modules in order to minimise the number of electrical interconnections and reduce overall mounting costs. In response PV manufacturers have steadily increased module sizes and power ratings that, in the case of silicon, now range up to several hundred peak watts. However such advances must be weighed against the difficulty of handling the largest, and therefore heaviest, modules.

Figure 3.2 shows a cut-away view of the edge of a typical module containing crystalline silicon solar cells. The cells, which are brittle, are cushioned by encapsulation in an airtight layer of *ethyl vinyl acetate* (*EVA*) to ensure that they survive handling. On top is a cover of tempered glass which is sometimes treated with an antireflection coating (ARC) to maximise light transmission. Underneath is a sheet of *Tedlar*, a light synthetic

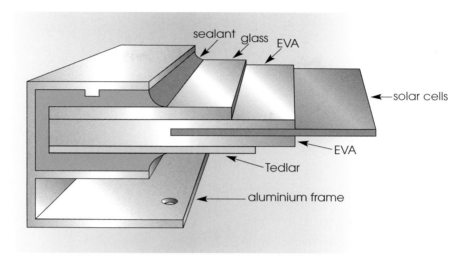

Figure 3.2 Typical construction of a conventional PV module.

polymer, acting as a barrier to moisture and chemical attack. The whole 'sandwich' is located in a slot in the aluminium frame and fixed with sealant. This construction must withstand up to 25–30 years of outside exposure in a variety of climates that include desert sands, alpine snows, wind, rain, pollutants, and extremes of temperature and humidity – a highly demanding specification. When things go wrong, it is often due to ingress of moisture or corrosion of electrical contacts rather than faults in the solar cells.

PV module design is by no means static, especially with the new thin-film technologies. In most cases the films, deposited on glass or other substrates, are scribed to produce the complete pattern of solar cells and interconnections, avoiding the need to handle and mount individual semiconductor wafers. Not only does this reduce manufacturing costs, it also promises improved electrical reliability within the module. Another area where thin films are having a big impact is in flexible products. The historic market dominance of rigid, relatively heavy, glass-covered modules is increasingly challenged by flexible lightweight designs more easily tailored to the shapes of awkward roofs and unusual structures or the aesthetic demands of architects. With modules as with solar cells, PV is in a phase of exciting and rapid development.

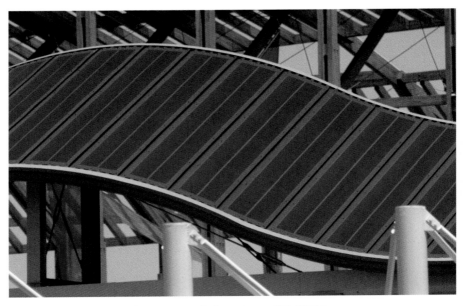

Figure 3.3 Innovative design: an example from Japan (IEA-PVPS).

3.2 Electrical performance

3.2.1 Connecting cells and modules

Individual solar cells are hardly ever used on their own. A cell is essentially a low-voltage, high-current device with a typical open-circuit voltage of around 0.5 V, far lower than the operating voltage of most electrical loads and systems. So it is normal for a PV module to contain many series-connected cells, raising the voltage to a more useful level. For example, many manufacturers offer modules with 36 crystalline silicon cells connected in series, suitable for charging 12 V batteries. These modules have an open-circuit voltage V_{oc} of around 20 V and a voltage at the maximum power point V_{mp} of about 17 V, giving a good margin for battery charging, even in weak sunlight. As PV moves increasingly towards high-power grid-connected systems, the trend is for more cells per module giving higher output voltages – for example, the modules previously shown in Figure 2.1 each contain 72 cells producing about 35 V at the maximum power point.

Of course the peak power of a module is also one of its key characteristics. The surface area of a monocrystalline silicon solar cell is limited by the diameter of the original ingot, which in turn restricts its power output. Many cells must therefore be interconnected to produce substantial module power: for example 72 cells, fitted into a module of about $1.5\,m^2$ area, can yield up to about $200\,W_p$. Multicrystalline silicon cells, being cut from large cast blocks of silicon, are less restricted in area; thin-film cells even less so. But the modules must still incorporate many cells to achieve useful voltage levels, and have a substantial surface area to give a reasonable power output.

What happens when many cells are connected in series? The answer would be very straightforward if all the cells were identical and exposed to the same strength of sunlight: with n cells in series, the module voltage would be n times the cell voltage and the module current would be the same as the cell current. But in practice cells are not identical. There are small manufacturing tolerances, occasional minor damage due to cracking, and small temperature differences, depending on where cells are located in the module. If a module becomes partially shaded by buildings or trees, some cells receive more sunlight than others. In all cases the module's output is limited by the cell with the lowest output – the 'weakest link in the chain'. The resulting loss of power is referred to as *mismatch loss.*

Small mismatch losses are to be expected in commercial modules and are covered by manufacturing tolerances. They need not normally concern us. But additional losses can easily be caused by shading, which should obviously be avoided where possible. The situation can worsen dramatically if one cell in a string becomes truly 'bad' and fails to generate current. It then acts as a load for the other cells and starts to dissipate substantial power, which can lead to breakdown in localised areas of its p–n junction. Severe local overheating occurs, possibly causing cracking, melting of solder, or damage to the encapsulating material. This is known as *hot-spot formation.*

The hot-spot condition is illustrated in Figure 3.4. At the top in part (a) is a string of n cells of which $n - 1$ are 'good' and one is 'bad'. The string is shown short-circuited, which is the worst-case scenario. Since the cells are in series, the current must be the same for all. But whereas the good cells happily generate a solar current I_L, the bad cell cannot do so and is forced into reverse bias. With the string short-circuited, the bad cell is subjected to the full voltage and power output of the good cells, leading to breakdown and hot-spot formation.

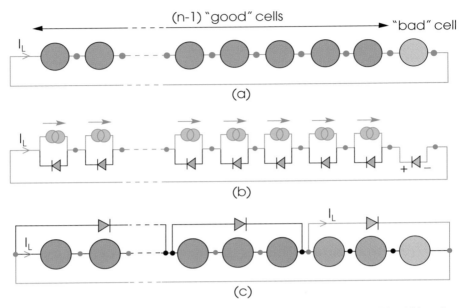

Figure 3.4 (a) A string of cells including one 'bad' cell; (b) equivalent circuit; (c) addition of bypass diodes.

Part (b) of the figure clarifies the situation from a circuit point of view. The complete current path is shown in red. Each good cell is represented by the simple equivalent circuit previously shown in Figure 2.9(a), consisting of a semiconductor diode in parallel with a current generator. Since the bad cell's current generator is inactive the circuit current I_L must pass through its diode in a reverse direction (of course, a diode is not supposed to pass reverse current; but in this case the voltage produced by all the good cells in series is sufficient to cause breakdown and make it conduct). The power produced by the good cells must now be absorbed by the bad cell since none is dissipated in an external load. In conditions other than short-circuit the situation is less severe, but hot-spots may still occur.

Hot-spot failure is avoided by incorporating additional diodes known as *bypass diodes*, shown in part (c). Here red indicates circuit elements carrying the main current, green indicates inactive (but serviceable) elements, and blue indicates the bad cell. The bypass diodes offer an easy current path around any bad cells. Ideally there would be one bypass diode for each solar cell, but this is rather expensive. Many PV modules therefore incorporate one diode per small group of cells. In our example there are three cells per group. The disadvantage is that each bad cell takes two good cells out of action, preventing them from contributing to the power output of the string. In practice the maximum number of cells in a group that prevents damage is normally reckoned about ten. Bypass diodes are often built into commercial modules but if they are not, care should be taken to avoid short-circuits, especially when there is partial shading by trees, buildings, or other structures.

Many of the ideas we have developed for solar cells also apply to modules. For example a module, like an individual cell, may be characterised by its open-circuit voltage, short-circuit current, and maximum power point (MPP). Indeed we may think of a complete module as a type of 'super-cell' with higher voltage and power ratings.

Modules are connected together – sometimes in large numbers – to form arrays. Whereas the cells in a single module are usually series-connected to raise the voltage as much as possible, modules in an array may be connected in series, parallel, or a mixture of the two. This is illustrated by the array of six modules shown in Figure 3.5, consisting of two parallel strings, each containing three modules in series. If the modules are perfectly matched this arrangement produces an array voltage three times the module voltage, an array current twice the module current, and an array power six times the module power. In practice the array performance will be slightly reduced by mismatch losses between the various modules. It is also worth noting that modules from different manufacturers should not be mixed together in an array, even if they are nominally similar, because differences in I–V characteristics and spectral response are likely to cause extra mismatch losses.

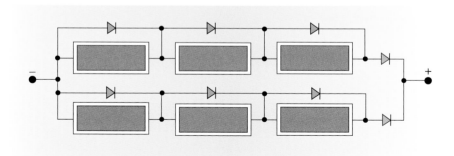

Figure 3.5 An array of 6 modules.

The figure also shows a number of diodes. Those coloured green are bypass diodes, one in parallel with each module, to provide a current path around the module if it fails or becomes 'bad'. The two diodes coloured red are referred to as *blocking diodes*, one in series with each string, to ensure that current only flows out of the modules. They are generally used in battery charging systems to prevent the batteries from discharging back through the modules at night. Some manufacturers include blocking diodes within their modules.

An array installed on a domestic roof might typically contain 10–20 modules. A single module is normally sufficient for the type of solar home system (SHS) that provides modest amounts of electricity for a family in a developing country. But the huge PV systems and power plants now being installed in industrialised countries incorporate hundreds of thousands of modules and involve major decisions about how they should be interconnected. A complete *photovoltaic generator* is, of course, a DC generator and since most large systems feed their solar electricity into an AC grid it is important to design arrays that can be interfaced safely and efficiently. We will return to this topic in the next chapter.

3.2.2 Module parameters

Not surprisingly, most of the electrical parameters of a PV module closely reflect those of its solar cells. However the efficiency of a module, measured in standard conditions of bright sunlight ($1000 \, \text{W/m}^2$ at $25 \, °\text{C}$, AM1.5 spectrum), is slightly less than that of the constituent cells because the cells do not completely fill the module's area and there are small power losses as sunlight passes through the top cover and encapsulant. If blocking diodes are built in, these too will produce small power losses. Typical efficiencies for the most widely used terrestrial modules are around 12–16% for monocrystalline silicon, 11–15% for multicrystalline silicon, 8–11% for CIS and CdTe, and 6–8% for *a*Si. But as we have emphasised previously, efficiencies in bright sunlight do not tell the whole story. Crystalline silicon modules tend to lose their advantage in weak or diffuse light, or high temperatures, and there is accumulating evidence that the newer thin-film modules may produce higher annual yields in regions with substantial cloud cover.

We have already described monocrystalline silicon solar cells in some detail and illustrated typical *I–V* characteristics for a $2 \, \text{W}_\text{p}$ cell at various levels of insolation in Figure 2.10(b). As we pointed out in the previous section, the cells in a module are normally series-connected, raising the power and voltage (but not the current). As an example we now consider a module containing 72 monocrystalline cells with a peak power rating of $180\,W_p$ – a popular size and power rating. The following module parameters, quoted by the manufacturer, are typical:

- Nominal power $180\,W_p$
- Open-circuit voltage   43.8 V
- Short-circuit current   5.50 A
- Voltage at maximum power   35.8 V
- Current at maximum power   5.03 A
- Power reduction per °C   0.45%
- Voltage reduction per °C   0.33%
- Length   1600 mm
- Width   804 mm
- Weight   18 kg
- Efficiency   14.0%

As expected, the nominal power of $180\,W_p$ equals the product of voltage and current at the maximum power point (MPP). The efficiency is given by the module power in kW_p divided by the module area in square metres. And the loss of power at elevated cell operating temperatures, 0.45% per °C, is typical of crystalline silicon – and more serious than for thin-film technologies. It means, for example, that if the cell temperature is allowed to rise to 65 °C the power output will fall by about 18%, emphasising the need to keep this type of module as cool as possible with adequate ventilation.

A family of $I\text{–}V$ curves for the module is shown in Figure 3.6. Their form is very similar to those for the $2\,W_p$ cell of Figure 2.10(b). The top curve, labelled $1000\,W/m^2$, refers to standard insolation and corresponds to the parameters in the above table. The other three curves confirm that as the level of insolation reduces, the current falls in proportion. Each has its own maximum power point (MPP), labelled $P_1\text{–}P_4$, the operating point at which maximum power output may be obtained.

Other aspects of module performance closely mirror those of the constituent solar cells. For example, since the module is effectively a current source, the actual power output is closely proportional to the voltage at which it is operated, up to the MPP; and the main effect of temperature rise on the $I\text{–}V$ curves is a reduction in open-circuit voltage. You may like to refer back to Figures 2.11 and 2.12 and the accompanying text for a discussion of these points.

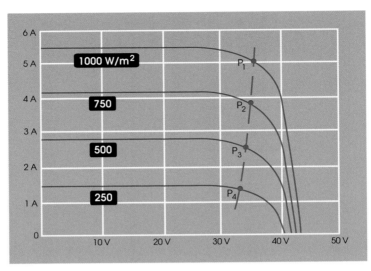

Figure 3.6 *I–V* characteristics of a typical monocrystalline silicon module rated at 180 W_p.

We have so far concentrated on monocrystalline silicon and you may be wondering how module parameters and *I–V* characteristics differ for other cell technologies. In fact the differences are very slight for multi-crystalline silicon, the main effect being a small decrease in module efficiency (and increase in module area) for a given power rating. The sizes and power ratings of both types of crystalline silicon modules have tended to rise steadily in recent years; most current production is in the range 150–300 W_p. In part this is due to advances in manufacturing processes, in part to satisfy a market increasingly slanted towards large grid-connected systems. The first commercial module to achieve 500 W_p appeared a few years ago.

When we come to the newer thin-film modules, the situation changes in two main respects. First, manufacturers generally start testing the market for new cell technologies with relatively small modules. It takes a lot of skill and experience to produce large modules with consistent performance and reliability. CIS and CdTe modules, for example, have worked up from small to medium sizes over a number of years, but maximum

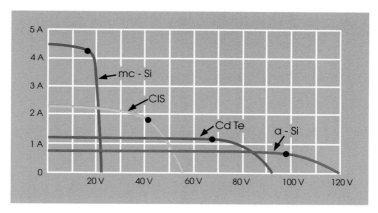

Figure 3.7 *I–V* characteristics in strong sunlight (1000 W/m²) of four 75 W$_p$ modules.

power ratings are still well behind those of crystalline silicon. Large a-Si modules, a much more mature technology, offer power levels into the hundreds of watts, but at amorphous silicon's relatively low efficiency. The second point is that the manufacture of thin-film cells offers considerable flexibility over module voltages. Cells can be scribed in a wide range of sizes within a module; a few large cells give high currents at low voltages, many small cells give high voltages at low currents. Higher voltages are often preferred for interfacing to an electricity grid and for reducing cable losses when modules are interconnected, a significant factor in large PV systems.

Figure 3.7 illustrates this flexibility with *I–V* characteristics for four commercial modules based on different technologies; multicrystalline silicon (mc-Si), CIS, CdTe, and a-Si. The modules all have the same nominal peak power of 75 W$_p$ but very different voltage and current levels. The curves are all drawn for standard insolation (1000 W/m²) and the maximum power point for each module is shown by a dot. The figure is meant to be indicative, not definitive, and the situation is of course developing rapidly. The main point is the increasing range of voltages levels offered by thin-film deposition and scribing techniques. However we should remember that the great majority of today's commercial PV modules, installed and in production, are still based on crystalline silicon and it will be many years before thin-film products dominate the scene.

83

3.3 Capturing sunlight

3.3.1 Sunshine and shadow

The cost-effectiveness of a PV system depends crucially on positioning its solar array to capture as much sunlight as possible. We must therefore appreciate how the Sun's apparent path across the sky varies according to the time of year and the latitude of the site. Some major features of the solar trajectory are illustrated in Figure 3.8 for sites in the northern hemisphere. The trajectory is lowest at the *winter solstice*, around December 21, and highest at the *summer solstice*, around June 21. In between are the two mid-season *equinoxes* around March 21 and September 21, when the Sun rises due east, sets due west, and gives equal hours of day and night. The high point is always reached at *solar noon* when the Sun is in the south. In winter, sunrise occurs south of east and sunset occurs south of west; in summer both veer towards the north. In other words the angular (azimuth) span as well as the height of the trajectory varies with the season. This all applies equally well to the southern hemisphere if we swap north for south, and interchange the dates of the winter and summer solstices.

Time measured by the apparent motion of the Sun is called *solar time* and fluctuates slightly around the time given by a conventional clock. This is

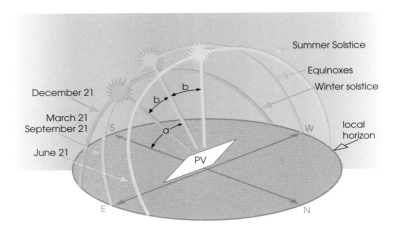

Figure 3.8 Solar trajectories.

because the Earth's journey around the Sun is slightly elliptical, and our distance to the Sun varies with the time of year. In this chapter we always refer to solar time (for example, *solar noon*) because we are, of course, interested in the apparent movement of the Sun in relation to PV installations. However the time system used in our everyday lives and shown by our clocks and watches averages out the fluctuations to make every day of equal length, and is referred to as *mean time*. The best-known example is *Greenwich mean time*, being the time measured at the Greenwich Observatory in London. This is on the *prime meridian*, 0° longitude, literally 'where east meets west'. Fortunately the difference between local solar time and mean time, described by the so-called *equation of time*, never exceeds about 17 minutes at any time of year. This is only really significant when designing PV systems that use highly concentrated sunlight and must track the Sun very accurately across the sky. We shall meet them later in the chapter.

Latitude also has a big effect – the further we are from the equator, the lower the Sun's path through the sky. On the equinox dates its elevation angle above the horizon at solar noon, labelled *a* in the figure, is equal to 90° minus the latitude. For example in Madrid, latitude 40°N, the noon elevation on March 21 and September 21 is 50°; in more northerly Berlin, latitude 52°N, it is 38° (and at the North Pole, latitude 90°N, the Sun is on the horizon). At the summer solstice, June 21, the noon elevation increases by an angle *b* equal to 23.45° and is at its annual peak. At the winter solstice it is reduced by the same amount. So in Madrid the summer and winter solstice elevations are 73.45° and 26.55° respectively (intrepid explorers at the North Pole for the winter solstice, in total darkness, are perhaps unaware that the Sun is 23.45° *below* the horizon!). These seasonal variations are caused by the offset angle between Planet Earth's axis of rotation and its plane of revolution around the Sun.

When positioning a PV array it is very important to avoid shadows as far as possible, for two main reasons. Shading can greatly reduce the output of the modules; and in severe cases it runs the risk of hot-spot formation. What may be termed 'occasional shadows' caused by bird droppings, dust layers, or snow on PV modules can obviously be reduced by proper maintenance and cleaning. But 'recurring shadows', due to local features, are more awkward. The degree of shading at different times of year depends upon the Sun's trajectory and may be assessed by recasting Figure 3.8 in two-dimensional form and adding the outlines of buildings, trees, and high terrain that threaten to cast shadows over the PV array. This is illustrated in Figure 3.9 for latitude 40°N, relevant to world cities including Madrid,

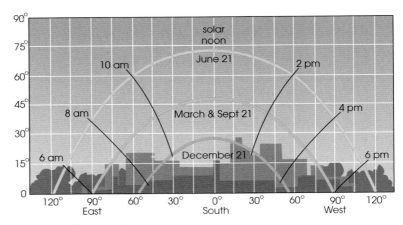

Figure 3.9 Shading effects.

New York, Ankara, and Beijing, and shows clearly the different amounts of shading in the summer and winter months. More detailed assessments can be made using a dedicated computer program such as the one published by the University of Oregon.[1] There are also a number of devices on the market for predicting potential shading problems, ranging from simple hand-held viewers to sophisticated photographic instruments supported by computer software.

The situation becomes more complicated if shadows are cast by nearby obstructions. For example PV roofs may be partially shaded by dormer windows, satellite dishes, chimneys or ventilation pipes; a small, ill-positioned pipe at two metres can cause more trouble than a skyscraper at two kilometres! Of course small local obstructions should be easier to control, and perhaps eliminate. A newly designed roof should always take special care to avoid them.

When shading is unavoidable it may be possible to reduce its effects by careful planning of module interconnections in the PV array. As we noted in the previous section, a single 'bad' or shaded cell in a series-connected string affects all the other cells and can seriously reduce the string's output. The same applies to strings of modules. So it is important to try and prevent one module in a string from becoming shaded at the expense of the others. And if a shadow is large enough to fall on several modules at the same time, it is best if all are members of the same string.

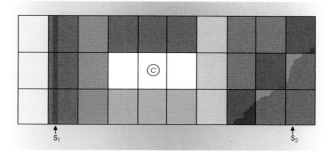

Figure 3.10 Arranging module strings to reduce the effects of shading.

These ideas are illustrated in Figure 3.10 for a PV roof containing 30 modules arranged as 10 parallel strings of 3 modules each. The various strings, indicated by different green tints, have been arranged to take account of two shadows, S_1 and S_2. The first of these is a narrow stripe formed by a nearby electricity pole, and its effects are reduced by a vertical arrangement of modules. As the shadow moves laterally in line with the Sun, it is mainly confined to a single string and affects three modules equally. The second shadow, cast by a neighbouring tree, is roughly triangular in shape and falls on the lower right-hand corner of the array. Assuming trees and neighbours are to remain undisturbed, modules may be connected in triangular strings. Once again this minimises the number of strings affected by the shadow as it moves onto the corner of the roof. And finally there is an unfortunate chimney pipe, labelled *c,* near the middle of the roof that cannot be moved. Its nuisance value is reduce by a string of 'dummy' modules, shown white, which preserves the array's appearance, but avoids using expensive real modules that would produce little electricity. In practice it would probably be economical to connect all the unshaded modules in one or two longer strings, a matter we shall return to later.

Although such 'array design' is only partly effective, it is virtually cost-free – an important benefit since recurring shadows can degrade an array's output over its entire working life. Finally, it is worth noting that a number of manufacturers now offer *power optimisers* that aim to reduce shading and mismatch losses by allowing each module in a string to operate at its MPP, regardless of what the others are doing. In principle such devices should be able to overcome many of the shading problems discussed above. But, of course, they come at a cost.

3.3.2 Aligning the array

In the previous section we saw how the Sun's trajectory varies according to the time of day, the season, and the latitude. This information suggests how a fixed PV array should be aligned to capture as much sunshine as possible. First, it should point due south towards the midday Sun. In some cases, for example on existing buildings, this may not be possible, but any deviation from south should preferably not exceed about 30°.

Second, the array should be tilted down from the horizontal so that the Sun's rays at solar noon are normal to its surface. Since the Sun's noon elevation varies continuously through the year, a choice has to be made about when to meet this condition. Very often the two equinoxes (about March 21 and September 21) are selected, giving the geometry shown in Figure 3.11. On these two dates the array points 'perfectly' at the midday Sun, but is somewhat too low in the summer and too high in the winter – normally a good compromise. In the previous section we noted that the Sun's noon elevation at the equinoxes (denoted by angle a in the figure) equals 90° minus the latitude of the site. It follows that its *declination* is equal to the latitude and that the array must be tilted down by this amount. For example in Madrid, latitude 40°N, an array must be tilted down 40° to meet the above condition. It will then point too high by angle b (23.45°) at the winter solstice and the same amount too low at the summer solstice.

The above 'equinox criterion' for tilting a PV array is widely adopted but, as we shall see, it is not essential. Minor variations of tilt angle have very little effect on an array's annual yield, and in any case there are some situations where a different choice of tilt may prove beneficial. A good example is a stand-alone PV system in a high latitude, required to provide a steady supply of electricity throughout the year. The winter months are the most difficult and will determine the size of array required, for if the system is able to cope in the winter it will certainly do so in the summer. So the downward tilt of the array is often increased to make the most of winter sunshine. Unusual climatic conditions may also favour different amounts of tilt, for example in parts of South Asia typified by hot humid summers with overcast skies followed by clear cool winters with plenty of sunshine. If it is required to maximise the annual electricity yield of a grid-connected system, a larger downward tilt may well be helpful. Conversely, in a system required to optimise electricity yield in the summer months, the PV array may be aligned closer to the horizontal. A good example is the summer holiday home described in Section 5.4.1.

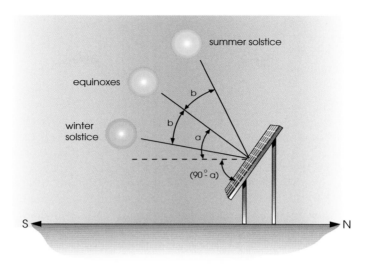

Figure 3.11 Aligning a PV array.

In the case of rooftop arrays, tilt is often predetermined by roof geometry, giving little or no flexibility (Figure 2.1 has already illustrated a large Swiss PV roof with a downward tilt rather different from the latitude of the site!). However it is worth noting that buildings in high-latitude countries such as Norway and Sweden often have high-tilt roofs to encourage snow to slide off easily; whereas roofs in Morocco or Egypt are much more likely to be flat, or nearly so. In this way vernacular architecture tends to suit the Sun's trajectory and the preferences of PV system designers.

So far we have concentrated on capturing as much of the Sun's direct radiation as possible. This is certainly important but, as our discussion of the solar resource in Section 1.2 made clear, there is also diffuse and albedo radiation to consider (see Figure 1.7). Indeed the only PV systems that rely solely on direct sunlight are those launched into space. What happens when we come down to Earth and start considering the actual radiation falling on a PV array, taking into account scattered light? And how much of the precious 'fuel' can actually be converted into electricity?

You may find it helpful to refer back to Figure 1.5 showing the large-scale effects of climate on insolation at the Earth's surface. In temperate regions with plenty of 'cloudy-bright' weather the diffuse component can make a surprisingly large contribution to the annual total; for example in Western Europe it is often over 50%. One effect is to make array alignment less critical because diffuse light tends to come from all over the sky. Another is to reduce the overall importance of shadows in determining the annual energy yield.

To narrow these general ideas down to a particular PV system we need more detailed local information. Fortunately, the great surge of interest in solar energy in recent years has spawned data on average sunlight conditions for many cities and locations around the world (one valuable source of information is provided by NASA.[2]) The data is often presented in the form of 12 monthly mean values of global (direct and diffuse) daily radiation on a horizontal surface, expressed in kWh/m^2. Albedo radiation is not included since it does not affect horizontal surfaces and anyway is highly site-dependent. Sometimes the proportions of direct and diffuse light are found by practical measurements with specialised instruments; sometimes they are inferred from the global figure and a *clearness index* summarising the amount of light scattering caused by clouds and particles in the local climate. Figure 3.12(a) shows a typical distribution for a West European city such as London or Amsterdam with a temperate climate giving plenty of 'sunshine and showers' in summer and cloudy skies in winter. The height of each bar represents global radiation, composed of direct (yellow) and diffuse (orange) components. The daily average over the whole year is about $2.8\,kWh/m^2$, giving an annual total of about $1050\,kWh/m^2$. Part (b) of the figure is for the Sahara Desert. Here the extremely sunny, hot and reliable climate produces a daily average of about $6\,kWh/m^2$ and an annual total of about $2200\,kWh/m^2$. Most of the radiation is direct.

Given such figures it is quite easy to make a rough estimate of the annual output from a PV module or array using the concept of *peak sun hours*. If the energy received throughout the year is compressed into an equivalent duration of standard 'bright sunshine' ($1\,kW/m^2$), then the number of peak sun hours is the same as the global annual figure. For example London has about 1050 peak sun hours in a year, so a PV module rated at $200\,W_p$ can be expected to produce around $1050 \times 200 = 210\,000\,Wh = 210\,kWh$ per year. However this is for a horizontally mounted module – unlikely in London. It also assumes ideal 'bright sunshine' conditions, whereas much of London's sunlight is diffuse. Direct and diffuse light have different spectral distributions and solar cells do not generally respond equally to them; nor are most cells equally efficient in bright and low-level light. So estimates of array output based on peak sun hours are only very approximate, especially for locations with a large proportion of diffuse light. In

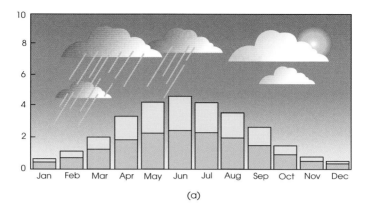

(a)

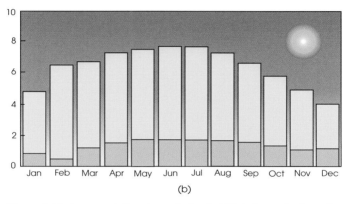

(b)

Figure 3.12 Average daily solar radiation in kWh/m² on a horizontal surface: in (a) London or Amsterdam; (b) in the Sahara Desert.

most cases actual annual yield is considerably lower. For example our 200 W$_p$ module in London is more likely to produce 150–160 kWh per year.

It must be emphasised that distributions such as those in Figure 3.12 are normally based on data collected over many years. In a given year, especially in unpredictable climates, they may look very different; it is not unusual to see a 10% variation in annual figures or a 30% variation in monthly ones, and this must be taken into account when making predictions or designing a PV system. There is also climate change to consider. If, as many people believe, we are now entering an era of extreme weather events and disruptions to traditional weather patterns, the accuracy of predictions based on averaged historical data becomes questionable.

91

We have so far considered sunlight falling on a horizontal surface. What happens when a PV array is tilted downwards to take account of the latitude? How are the figures for global, direct, and diffuse radiation affected? These are complicated matters so we just give a short summary, referring you to more advanced books for detailed explanations.[3–5]

Predicting solar radiation on an inclined (tilted) south-facing PV array involves the following steps:

- As above, obtain data or estimates of direct and diffuse radiation on a horizontal surface for the particular location.

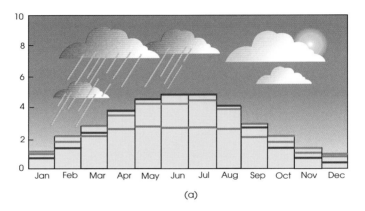

(a)

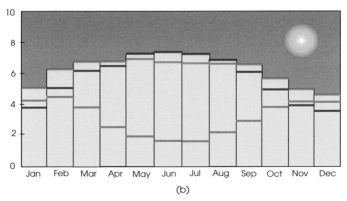

(b)

Figure 3.13 Daily solar radiation in kWh/m² on south-facing inclined PV arrays in: (a) London; (b) the Sahara Desert. In each case three values of tilt are illustrated: 0° (blue), the latitude angle (red), and 90° (green).

- Using the clearness index and empirical formulae, adjust the data to account for the inclined surface. It is normally assumed that diffuse radiation comes equally from all parts of the sky and is therefore unaffected by the amount of tilt.

- Where appropriate, estimate the albedo contribution using reflectivity values for typical ground surfaces. In many cases the albedo is small or insignificant, but snow cover can be particularly relevant.

Fortunately, various government research institutes and universities publish programs that perform the necessary computations, the information provided by NASA[2] being particularly comprehensive. Figure 3.13 shows some estimates for south-facing tilted PV panels in London (latitude 52°N), and the Sahara Desert (latitude 24°N). In each case three different tilt angles are illustrated: 0° (horizontal) shown by blue bars; an angle equal to the latitude, shown by red bars; and 90° (vertical), shown green. You may be surprised at the choice of 0°and 90°, but actually results for angles closer

Figure 3.14 Vertical and in diffuse light: a large PV façade in Manchester, England (IEA-PVPS).

93

to the latitude are often almost indistinguishable. And remember that horizontal PV arrays may be installed on flat roofs, and vertical ones on building façades! The results illustrate several interesting points:

- London: the results for 0° and 52° tilt are quite similar, mainly due to the large proportion of diffuse sunlight, but 0° receives slightly more radiation in the summer and less in the winter; 90° tilt (as on a vertical building façade) has a big effect on radiation in the summer months.

- Sahara Desert: the results for 0° and 24° tilt are very similar since we are much closer to the equator. The really big effect is the reduction in radiation for a 90° tilt in summer, when the Sun is high in the sky and the radiation is almost all direct.

Of course such graphs are estimates that cannot take account of fine variations in local climate – for example, the different amounts of cloud and shade on opposite sides of a valley. As PV enters the multi-gigawatt era, with systems of all shapes and sizes installed around the world, system designers will no doubt have access to ever-more performance data collected from working systems.

3.4 Concentration and tracking

Ever since the dawn of the modern photovoltaic age the PV community has pondered the attractions of concentrated sunlight. After all, if the price of solar cells is very high (and it certainly was in the early days!) and is closely related to their surface area, it should make sense to focus the Sun's light onto cells of very small area. Furthermore, specialised cells designed to work under concentrated sunlight can achieve considerably higher conversion efficiencies than conventional cells. For example, in our discussion of gallium arsenide cells in Section 2.4.3.1 we noted that efficiencies around 40% make them suitable candidates for high-concentration PV systems. But the approach is only viable if efficiency improvements and cost savings on the cells more than offset the additional costs of lenses or mirrors plus, in most cases, equipment to track the Sun on its daily journey across the sky. Unsurprisingly, there are plenty of sceptics, not least because the cost of conventional solar cells and modules continues to fall. But the jury is still out, and in recent years remarkable progress has been made in the design and production of high-performance PV concentrator systems. It will be fascinating to see how the market develops in the coming decade.

To summarise, such systems offer two main attractions:

- The area of the solar cells can be greatly reduced.
- Cells designed for high-intensity concentrated sunlight can achieve better conversion efficiencies than standard cells.

However the disadvantages and challenges appear rather onerous:

- Lenses or mirrors must be used to concentrate the light.
- Above a certain level of concentration it becomes essential to track the Sun across the sky, keeping the focused light accurately aligned on the solar cells.
- High concentration is effective for the direct component of sunlight, but not the diffuse and albedo components.
- Focusing and tracking equipment must be robust and properly maintained to match the expected lifetime – say 25 years – of solar cells and modules.
- Tracking systems are generally unsuitable for building-integrated PV, including rooftop arrays.

We may therefore expect to see high-concentration tracking systems maintained by professional staff and largely restricted to power plants in areas with a high percentage of direct sunlight. Low-concentration systems, which do not need tracking, are more likely to find favour for rooftop and other static installations.

So how is sunlight concentrated and focused onto small-area solar cells? There are two basic approaches: using transparent lenses, or reflective mirrors. The first of these is illustrated in Figure 3.15. Part (a) shows a

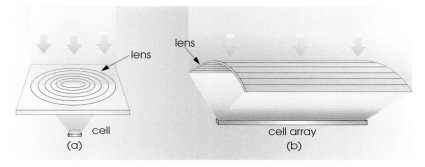

Figure 3.15 Concentrating sunlight onto solar cells using lenses: (a) a circular Fresnel lens with point-focus; (b) a linear Fresnel lens with line-focus.

Figure 3.16 Two-stage focusing of light to achieve high concentration (EPIA/Isofoton).

circular lens, normally made of plastic, which concentrates the direct sun-light onto a small solar cell. Simple refractive lenses become very thick if their diameters exceed about 10 cm, so a special form known as the *Fresnel lens* is widely used. Rather than allowing the lens to get thicker and thicker towards its centre, the convex surface is collapsed back to a thinner profile in a series of steps. A family of these lenses, each focusing sunlight onto a single solar cell, can be built up as a parquet to make a large flat PV module.

Whereas the circular Fresnel lens is *point-focus*, the linear, domed, form of Fresnel lens shown in Part (b) of the figure produces a *line-focus* onto a long array of cells. Once again the lens profile is collapsed in a series of steps, keeping its thickness reasonably constant around the curve. The curvature increases mechanical strength and avoids optical problems that can arise with more flexible, flat, lenses.

In high-concentration systems the necessary focussing is sometimes achieved in two stages. The main lens performs an initial concentration of

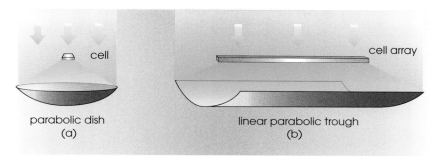

Figure 3.17 Concentrating sunlight with reflective mirrors.

the light, directing it onto a *secondary optical element* for further concentration. This also offers an opportunity to ensure that the intensity of light striking the active area of the solar cell is as uniform as possible.

Reflective mirrors provide an alternative to lenses. You are probably aware that a parabolic dish mirror receiving light parallel to its axis brings the light to a point at its focus. This is shown in Figure 3.17(a) with a solar cell mounted at the focus. Another effective configuration is the linear parabolic trough shown in part (b) that focuses the incoming light onto a linear array of cells.

The degree of concentration achieved by a lens or mirror is commonly expressed in *suns*. This is the ratio between the intensity of the incoming sunlight, normally taken as the standard insolation of $1000 \, W/m^2$, and the average intensity of the light focused onto the active area of the solar cell, or cells. For example a concentration of 100 suns produces a nominal $100 \, kW/m^2$ or $10 \, W/cm^2$ at the cell surface. Note, however, that in practice the insolation of $1000 \, W/m^2$ is not all direct sunlight, even under clear skies. Typically 85% is direct, the other 15% diffuse. So a nominal 100-sun concentrator would more likely produce about $8.5 \, W/cm^2$ at the cell surface in strong sunlight – and systems are sometimes rated on this basis. The amount of concentration in practical systems varies from as little as 2 or 3 suns in static systems that do not need to track the Sun up to 1000 suns in high-concentration tracking systems, some of which now employ multi-junction gallium arsenide solar cells.

Unlike solar cells used in conventional PV modules that must be illuminated over their entire area for efficient performance, small concentrator cells are often designed with an 'active area' surrounded by a nonilluminated edge carrying bus bars and connections. This means that the reduction in cell area, one of the main advantages of a concentrator, is less than its number of suns. Furthermore, when many small cells are cut from a semiconductor wafer, there is quite a lot of wastage at the edges. The net reduction in wafer usage is substantially less than might be expected from a simple consideration of the concentration ratio.

Two more aspects of concentration optics should be mentioned briefly. The first of these is *acceptance angle,* the angular range over which a concentrator can accept light from the Sun. Clearly, in the case of a tracking system the greater the acceptance angle the better, because it minimises the tracking accuracy required and hence the complexity and cost of the tracking equipment. But perhaps not surprisingly there is a fundamental trade-off between acceptance angle and concentration: increasing the acceptance angle reduces the amount of concentration attainable, and *vice versa.* An engineering compromise is required.

The second aspect is *nonimaging optics.* As any student who has played with a magnifying glass knows, light from distant objects can be brought to a focus on a sheet of white paper, producing an inverted image of the

Figure 3.18 The *Euclides* 480 kW$_p$ system on the island of Tenerife (Spain) comprises a series of 14 linear parabolic trough mirrors 84 m long (EPIA/BP Solar).

scene. If the magnifying glass is used to focus the Sun's rays onto the paper in an attempt to set it alight, the brilliant circular dot of light is an image of the Sun. These are examples of classical *imaging optics*. But in the case of PV concentrators there is no particular virtue in obtaining an image of the Sun. It is more important to illuminate the active area of the solar cell, or cells, as uniformly as possible using a lens or mirror system with as large an acceptance angle as possible. Such design considerations have led to big advances in nonimaging optics applied to PV systems.[6]

It is now time to turn our attention to tracking. It is often stated that PV modules that track the Sun deliver about 40% more electricity in an average year than static modules. Although the figure is a useful guide, the actual benefits of tracking depend on the local climate and the degree of concentration. We can summarise the situation with a few key points:

- Many systems, including large power plants, are based on flat-plate PV modules and use tracking without any concentration.

- Conversely it is possible to use a certain amount of concentration – say up to 6 or 8 suns – without any tracking. Static concentrators are attractive in principle because the costs and complexities of tracking equipment are avoided. But the need for optics makes it difficult for them to compete with conventional modules, especially as the price of thin-film solar cells continues to fall.

- A very basic form of tracking can be achieved manually. By adjusting the orientation of a flat-plate or low-concentration module just three or four times a day, in line with the Sun's trajectory, over 90% of the electricity yielded by a fully automatic tracker may be obtained. This is an interesting possibility for small systems with just one or two modules, for example solar home systems in developing countries where family members are generally on site and can easily make the adjustments.

- Moving to automatic tracking, this can be around either one or two axes. One-axis tracking is generally adequate for nonconcentrating systems and for systems using low-to-medium concentrations (up to about 40 suns), but two-axis tracking becomes necessary for high-concentration systems in locations with a high percentage of direct sunlight. The cost and complexity of accurate two-axis tracking make it suitable for power plants employing professional maintenance staff.

Two common types of one-axis tracking are illustrated in Figure 3.19. In part (a) a linear parabolic trough mirror rotates about a horizontal axis,

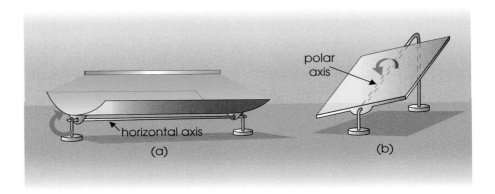

Figure 3.19 One-axis tracking.

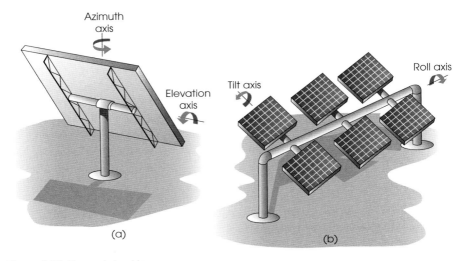

Figure 3.20 Two-axis tracking.

oriented either north–south or east–west. In either case optimal focusing is only achieved when the mirror points directly at the Sun; at other times of day the line image moves along the focus axis causing some end losses and image broadening. The resulting reduction in efficiency is offset by the relative simplicity of the tracking scheme, which is also very economical in its use of ground area – indeed hardly any more is needed than with static modules. The system illustrated in Figure 3.18 is also of this type.

In Figure 3.19(b) tracking takes place about a polar axis aligned with the Earth's axis of rotation. This limits the Sun's offset angle to a maximum of 23.45° from the plane of illumination, giving more efficient overall energy collection than with a horizontal axis (for an explanation of the angle you may like to refer back to Section 3.3.1). However the high profile means that more ground area is required to avoid shading of adjacent trackers, wind loading tends to be more serious, and the mounting is more awkward.

High-concentration optics generally demand tracking about two axes so that the focused light always falls accurately on the solar cells. Two schemes are illustrated in Figure 3.20. Part (a) shows the widely used pedestal form of tracker with rotation about a horizontal (elevation) axis and also a vertical (azimuth) axis. This scheme is simple to install, but tends to suffer from high wind loads, producing large torques on the drive system. Large trackers with surface areas up to $250\,m^2$ or more normally adopt a horizontal

Figure 3.21 This impressive two-axis tracker in Las Vegas, USA, supports multiple point-focus concentrator modules housing multijunction GaAs solar cells and is rated at $53\,kW_p$ (Amonix Inc.).

position in very high winds. Part (b) shows a less common form, known as the roll-and-tilt tracker. Wind loading is generally less serious, but more bearings and supports are needed.

In moving from static systems to one-axis, and then two-axis, trackers it is clear that capital and maintenance costs of mechanical and drive components must increase markedly. The challenge for designers of high-concentration systems is to engineer products with high technical performance and reliability that, in suitable climates, can rival the overall costs of systems based on more conventional approaches.

References

1. University of Oregon (solardat.uoregon.edu). *Sun Chart Program, Solar Radiation Monitoring Laboratory* (2010).
2. NASA (eosweb.larc.nasa.gov/sse). *Surface meteorology and Solar Energy Tables* (2010).
3. T. Markvart (ed.). *Solar Electricity*, 2nd edn, John Wiley & Sons, Ltd: Chichester (2000).
4. S.R. Wenham *et al. Applied Photovoltaics*, Earthscan: London (2007).
5. F. Antony *et al. Photovoltaics for Professionals*, Earthscan: London (2007).
6. A. Luque. *Solar Cells and Optics for Photovoltaic Concentration*, Adam Hilger: Bristol, Philadelphia (1989).

4 Grid-connected PV systems

4.1 Introductory

PV systems are generally divided into two major categories: *grid-connected* (also known as *grid-tied*) systems that are interfaced to an electricity grid; and *stand-alone* systems that are self-contained. Over the years it has been customary for books on PV to describe stand-alone systems first, probably because they are seen as 'pure PV'. Also we should remember that stand-alone systems, including those launched into space and the solar home systems (SHSs) that supply electricity to individual families in developing countries, accounted for much of the PV industry in its early days. But since the 1990s the market has shifted decisively towards PV power plants and installations on buildings connected to an electricity grid. By the year 2000 grid-connected PV had overtaken stand-alone systems in global market share and by 2009 more than 95% of solar cell production was being deployed in grid-connected systems. In many ways such systems are simpler to design and describe than their stand-alone cousins. For both these reasons our own story begins with grid-connected PV.

Since most people have seen PV arrays mounted on the roofs of homes, this seems a good place to start. Figure 4.1 shows the elements of a domestic PV installation, typically with an array power between 1 and $5\,kW_p$, interfaced to the local electricity grid. The major advantage of this arrangement is that the output from the PV array is fed into the grid when not required in the home; conversely, when the home needs power that cannot

Electricity from Sunlight   By Paul A. Lynn
© 2010 John Wiley & Sons, Ltd

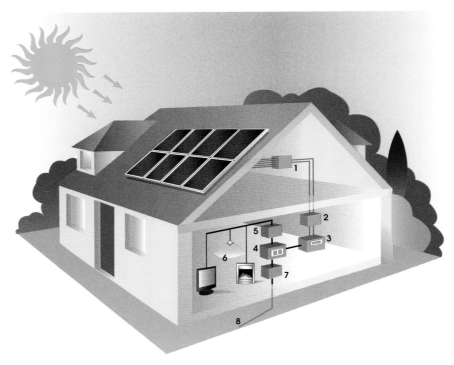

Figure 4.1 Connecting to the grid.

be provided by the PV (especially at night!) it is imported from the grid. In other words the PV system and grid act in harmony and there is an automatic, seamless back-and-forth flow of electricity according to sunlight conditions and the electricity demand.

In more detail the various items numbered 1–8 in the figure have the following functions:

1. *PV combiner unit.* This acts as a junction box connecting the modules in the desired configuration.

2. *Protection unit.* This unit houses a DC switch to isolate the PV array and antisurge devices to protect against lightning. Alternatively, these functions may be incorporated into units 1 or 3.

3. *Inverter.* At the heart of the grid-connection system, the inverter extracts as much DC power as possible from the PV array and converts it into AC power at the right voltage and frequency for feeding

into the grid or supplying domestic loads (an inverter may be thought of as the opposite of a rectifier that converts AC to DC).

4. *Energy-flow metering.* Kilowatt hour (kWh) meters are used to record the flow of electricity to and from the grid.

5. *Fuse box.* This is the normal type of fuse box provided with a domestic electricity supply.

6. *Electrical loads.* Domestic electrical loads include lighting, TVs, and heaters.

7 and 8. A junction box connects the home to the utility supply cable.

The adoption of domestic rooftop installations is mushrooming in developed countries in response to the falling prices of PV modules, the support of governments, and the enthusiasm of citizens to do something positive about global warming. Larger grid-connected systems, for example those installed on schools, offices, public buildings, and factories, extend the power scale up to hundreds of kilowatts or more. All have the advantage of generating solar electricity where it is needed, reducing the losses associated with lengthy transmission lines and cables. And at the top of the grid-connected power scale come multi-megawatt power plants, generally remote from individual consumers, which send all their power to the grid.

4.2   From DC to AC

The *inverter* is the key item of equipment for converting DC electricity produced by a PV array into AC suitable for feeding into a power grid. Inverters use advanced electronics to produce AC power at the right frequency and voltage to match the grid supply. While a single inverter may well be sufficient for a domestic installation such as that illustrated in Figure 4.1, multiple units become the norm as we advance up the power scale and their efficiency, reliability, and safety are major concerns of the system designer.

Inverters must obviously be able to handle the power output of a PV array over a wide range of sunlight conditions. Normally they do this using *maximum power point tracking* (*MPPT*) to optimise the energy yield. DC to AC conversion efficiencies up to 98% can be achieved over much of the range, although efficiency tends to fall off if an inverter is operated below about 25% of its maximum power rating. In larger systems with multiple inverters it can make sense to switch all the power into one unit at sunrise and then, as the Sun rises in the sky and the array power increases, bring

Figure 4.2 Raising the power level: a 17.6 kW_p grid-connected roof installation on the Oslo Innovation Centre, Norway (IEA-PVPS).

other inverters successively into play, keeping all working optimally. The switching sequence is reversed towards sunset. Overall, inverter system design is quite a challenge, especially with high-capacity units; few electronic systems are expected to maintain high efficiency over such a wide power range.

From the technical point of view there are two main classes of inverter: *self-commutated*, where the inverter's intrinsic electronics lock its output to the grid; and *line-commutated*, where the grid signal is sensed and used to achieve synchronisation. Inverters are also classified according to their mode of use, with four main types:

- *Central.* The complete output of an array is converted to AC and fed to the grid. The largest central inverters can exceed 1 MW_p capacity and weigh over 20 tons.

- *String.* This type of inverter is connected to a single string of modules with a typical power range of 1–3 kW_p. The weight is around 5 kg per kW_p.

Figure 4.3 This Korean power plant uses four 250 kW$_p$ inverters to connect 1 MW$_p$ of PV arrays to the grid. The nonconcentrating modules are mounted on horizontal single-axis trackers (IEA-PVPS).

- ■ *Multi-String.* These inverters can accept power from a number of module strings with different peak powers, orientations, and perhaps shading, allowing each string to operate at its own maximum power point (MPP).
- ■ *Individual.* An increasing number of manufacturers offer PV modules with inverters attached, making each module its own AC power source.

Several factors influencing the choice of inverters for small and medium-size systems can be explained by referring to Figure 4.4. For simplicity we have shown arrays with just a few modules although most systems contain more – and some a great many more. The array in part (a) consists of two strings of three modules each. In this case all the modules are assumed to be of the same type and rating, with the same orientation and without shading, so the strings are paralleled in the combiner/protection

107

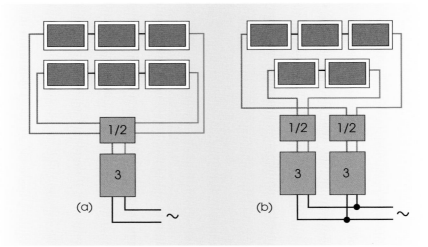

Figure 4.4 PV arrays served by: (a) a single central inverter; (b) two individual string inverters.

units (1/2) and fed to a single central inverter (3). The inverter is presented with an input voltage equal to three times the individual module voltage, and an input current equal to twice the individual module currents. Since the modules are well matched, the MPP selected by the inverter for the whole array ensures that all modules work at, or close to, their maximum output.

In part (b) of the figure the two strings are dissimilar. They may have different numbers of modules (as shown), or different module types or orientations; or one string may suffer partial shading. For whatever reason they do not produce similar outputs and cannot be efficiently characterised by a single MPP, so each string has its own inverter and is operated at its own MPP. An alternative is to use a single multi-string inverter. And as we have already pointed out in the previous chapter, manufacturers are now offering power optimisers, one to be connected to each module in a string, allowing every module to work at its own MPP. There are various options for extracting the maximum amount of power from strings and arrays.

To put our discussion in the context of a practical system, suppose we need to specify an inverter for a PV array of about $5\,kW_p$ on the roof of a suburban house. In a sunny climate an array of this size may well generate, over a complete year, electricity equal to the annual requirements of the household. In the summer months the PV will be a net exporter to the grid; in the winter months the solar deficit will be made up from the grid. We will assume that monocrystalline silicon modules rated at $180\,W_p$ have been selected, so 28 will be needed, yielding $5.04\,kW_p$ (the module specification given in Section 3.2.2 will be used in the calculations below). Fortunately, they can all be mounted on the roof at the same tilt angle and there is no shading, so we may specify a central inverter. Since an array rarely generates its nominal peak power, an inverter rated at slightly less than $5.04\,kW_p$ should be adequate so long as its maximum input voltage and current are never exceeded. We will therefore investigate the suitability of a $5\,kW_p$ central inverter with the following manufacturer's ratings:

- nominal DC input power: 5.0 kW
- peak instantaneous input power: 6.0 kW
- maximum DC input voltage: 750 V
- voltage range for MPP tracking: 250–650 V
- maximum DC current: 20 A

We first need to estimate how many modules can be connected in a series string. The maximum number is given by the maximum MPP tracking voltage of 650 V divided by the MPP voltage of an individual module. The latter is 35.8 V at 25 °C, but increases by 0.33% for every degree drop in temperature. Therefore if we allow for sunny winter days with temperatures down to −5 °C, the MPP voltage could reach 10% above 35.8 V, that is 39.4 V. The maximum number of modules in a string for effective tracking is therefore 650/39.4 = 16.5, say 16.

We should also check that the maximum DC input voltage of 750 V is never exceeded. Once again, the danger condition is a cold winter day with bright sunshine. The module open-circuit voltage of 43.8 V at 25°C rises by 10% to 48.2 V at −5 °C. So to keep within the 750 V limit the maximum number of modules is 750/48.2 = 15.6, say 15.

The minimum number of modules in a string is dictated by the need to keep the MPP tracking voltage above 250 V. The module's MPP voltage falls with rising module temperature, which could reach 70°C and cause a 15% drop in MPP voltage to 30.4 V. The minimum number of modules is therefore 250/30.4 = 8.22, say 9.

To keep within the inverter's voltage limits, we conclude that strings may have any number of modules between 9 and 15. Since the array contains 28 modules, two strings of 14 are acceptable, but not four strings of 7. Finally the array current supplied to the inverter should be checked to make sure it does not exceed the permitted maximum. In this case the peak, short-circuit, module current is 5.5 A, and is little affected by temperature. So two parallel strings of 14 modules will give a peak DC current of 11 A, well below the permitted maximum of 20 A. The inverter is therefore suitable for the job.

Back in Section 3.3.1 we discussed the problem of shading, and suggested reducing the effects of recurrent shadows by confining them to as few strings as possible. Where shading is unavoidable it may be appropriate to use a number of string inverters rather than a single central inverter, giving flexibility to connect the modules in a favourable configuration, perhaps with strings of different lengths.

We have already mentioned the need for inverters to operate efficiently over a wide power range. Some inverters include transformers, and these reduce efficiency slightly. High efficiency is not purely a question of economics; it also relates to keeping inverters cool. For example if a $5\,kW_p$ inverter is working at full stretch and converting 96% of its input power to AC, the other 4% (200 W) must be dissipated as heat. It is hardly surprising if the manufacturer recommends mounting the unit on an outside, north-facing wall with plenty of air circulation! The cooling issue becomes more and more significant as inverter power-handling capacity increases.

If the electricity grid is turned off for maintenance purposes, or due to a fault, it is very important for an inverter to disconnect itself automatically to avoid putting a voltage on the grid. Otherwise it can endanger personnel working on the grid, and may deceive other local inverters into believing that the grid is still operating normally. Sophisticated electronics are included to prevent this potentially hazardous situation, which is referred to as *islanding*.[1,2]

As we gaze at a domestic rooftop system rated at a few kW_p, it is hard to appreciate the engineering challenges posed by scaling up inverters for multi-megawatt power plants. There are major issues of technical performance to be considered including lightning and surge protection, safety, reliability, inverter sequencing, and the mode of connection of tens or hundreds of thousands of PV modules into strings and arrays.[1] The waveform purity and power factor of the inverter output must be satisfactory to the grid operator. Grid-connected inverters are sensitive to fluctuations in grid voltage, frequency, and impedance, and will shut down automatically if these parameters stray outside the agreed specification. Islanding, which could be disastrous in a large installation, must be avoided. All in all, high-power inverters provide a major challenge to today's electrical and electronic engineers.

Figure 4.5 Scaling up: this 1.6 MW$_p$ inverter weighs over 20 tons (Padcon GmbH).

Figure 4.6 The Moura power plant in Portugal, rated at 45.6 MW$_p$, represents a big challenge for inverters as well as PV cells and modules (IEA-PVPS).

111

4.3   Completing the system

Various items are required to complete a grid-connected PV system. They may be less glamorous than solar cells and PV modules, but they are essential to a properly engineered installation. Costs, long-term reliability, ease of maintenance, and sometimes appearance, are important considerations. They are generally referred to as *balance of system* (*BOS*) components.

As the prices of solar cells and modules continue to fall and PV manufacturers achieve the cherished long-term objective of 'one US dollar per watt', the cost of BOS components can, unless carefully controlled, seriously inflate total system costs. In the past a figure of about 50% has often been quoted, including inverters. One of the main problems has been a proliferation of components supplied by many manufacturers in small quantities, lacking the economic benefits of scale. As the PV industry continues to grow, there is perhaps a better chance that volume production will drive costs down.

We mentioned and illustrated various BOS components for a domestic PV installation in Section 4.1. It is now time to give a more complete list and add further comments:

Module and array mounting structures (Figure 4.7). Modules and arrays need secure mounting whether on the ground, flat roofs, inclined roofs, or building facades. A great variety of static mounting structures is available, in aluminium, stainless or galvanised steel. Some allow variable tilt. Generally there should be space left at the back of modules to allow free air circulation.

■ *Cabling.* Special double-insulated cables that are UV and water-resistant are generally used for the DC wiring from modules to inverters. They must be sized to give low voltage drops, typically less than 2%. Since cable power losses are proportional to the square of the current carried, there is an advantage in reducing current levels by specifying long module strings and high system voltages.

■ *PV combiner unit* (Figure 4.8). This acts as a junction box for the various module strings, which are normally connected in parallel. Fuses are provided for each string. The combiner box may include surge protection against lightning and house the main DC isolator switch – providing it is easily accessible – allowing the PV array to be disconnected from the inverter.

Figure 4.7 Array mountings at the Kings Canyon power plant in Australia (IEA-PVPS).

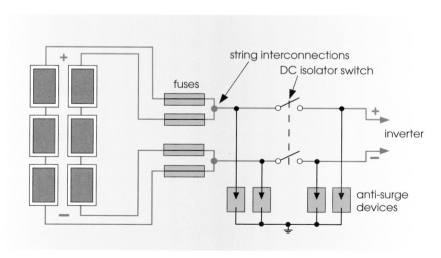

Figure 4.8 Key functions of the combiner/protection units in a domestic PV system.

113

- *Protection unit* (Figure 4.8). If the combiner unit does not include a DC isolator, this must be provided separately and be easily accessible. Since a PV array always produces a voltage in sunlight it must be possible to disconnect it from the inverter for maintenance or testing. The isolator switch is rated for the maximum DC voltage and current of the array. Other safety features, including earthing, vary from country to country and between continents, although regulations are tending to harmonise as the PV industry extends its global reach.

- *Energy-flow metering.* In Figure 4.1 we showed twin kWh meters recording the flow of electricity to and from the grid. An alternative is to use a single bidirectional meter to indicate the net amount of electricity taken from the grid. This approach, referred to as *net metering*, implies that electricity exported to the grid by the PV array achieves the same price as imported electricity, regardless of when it is generated. Net metering is, in this sense, beneficial to the homeowner; but it is not suitable for feed-in tariffs that offer an attractive price for exported kWh, or differential tariffs that price electricity according to the time of day or night. In any case, most homeowners wish to know how much electricity is being generated by their solar arrays and often choose to have a visual display fitted in a living area of the house. It forms a good talking point with visitors. In addition, most inverters incorporate data-logging facilities allowing the owner to monitor performance using a laptop, and some include wireless data transmission.

As we move up the power scale towards larger grid-connected systems the importance of accurate performance monitoring grows. Large PV power plants have a full range of instrumentation typical of modern, high-tech, industrial facilities. And their full complement of BOS subsystems and components account for a substantial part of overall costs.

4.4 Building-integrated photovoltaics (BIPV)

4.4.1 Engineering and architecture

We have seen how a grid-connected system is built up using PV modules, inverters, and BOS components. In previous chapters we included several photographs showing PV roofs and vertical facades. So what exactly is implied by the term *building-integrated photovoltaics* (*BIPV*), and what more is there to be said about giving buildings a 'face to the Sun'?

Photovoltaic technology is unique among the renewable energies in its interaction with the built environment. Future generations will find it entirely natural to see PV arrays on roofs and facades, in gardens and parks, on bus shelters and car ports, and as electricity-generating windows and screens inside homes, schools, offices, and public buildings. Most will be grid-connected. And hopefully they will bear testament to the trouble our generation has taken to blend them visually and aesthetically into their surroundings. PV will become increasingly a part of the urban experience.

By contrast, most wind power is generated in wild open country or offshore and whatever one thinks of the visual impact of large turbines, they rarely impinge on the urban and suburban scene. Wave and tidal power do not affect the daily visual experience of office workers or families – unless they go on trips to marvel at large-scale renewable energy in action. Large PV power plants may be impressive and even beautiful in their own way, but they are not generally noticed by city dwellers.

BIPV is different. It proclaims a message about our care for the environment, it can be anywhere and everywhere, and it matters what it looks like and how people feel about it. Public enthusiasm and support are vital, not least to the PV industry. Architects will, or should, be involved with engineers in the design of solar buildings so that PV is integrated into the fabric in ways that marry technical function with aesthetics. A modern factory producing solar cells, or an exhibition centre for renewable technologies, offers an ideal opportunity to create a striking building that makes a highly visible statement about our technological future; a family living in a low-energy timber-framed eco-home may see their PV modules as a symbol of sustainability, an alternative lifestyle. In their very different ways all wish to proclaim a message about renewable energy that can only be successfully communicated by high-quality BIPV.

Of course there are difficulties. Countries including England, France, Italy and Spain have a huge stock of old and historic buildings. It would be difficult or impossible to modify most of them to accept PV modules in aesthetically pleasing ways. PV arrays tacked on to existing roofs hardly ever increase their visual attraction. We may like to see them because of what they represent – the owner's commitment to renewable electricity – but our enthusiasm stems from what the PV does, not how it looks.

The main opportunity for successful BIPV, as opposed to PV that is simply superimposed on existing roofs and structures, lies in the creation of new buildings that, from the very start, treat PV as an integral part of the design,

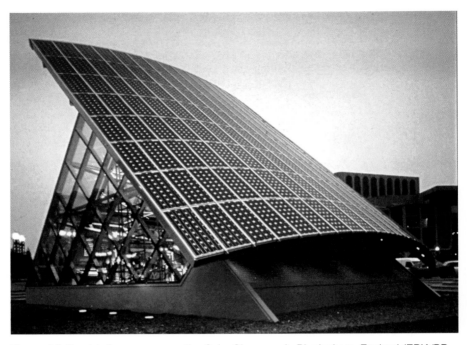

Figure 4.9 Proclaiming a message: the *Solar Showcase* in Birmingham, England (EPIA/BP Solar).

full of exciting possibilities. It is very encouraging to see architects in some countries – Germany and the Netherlands are good examples – realising that PV offers far more than a way of producing electricity. While appreciating its technical possibilities and limitations, their primary goal as architects is to ensure that PV enhances the human environment. For them PV is neither add-on nor afterthought, but an important part of the building and a pointer to its function and personality. It must inspire as well as serve its utilitarian purpose. As old housing and building stock is gradually replaced we may expect PV to exert a growing influence on architectural design, opening up hitherto undreamed-of possibilities.

Apart from aesthetics BIPV has several important economic advantages:

- In most cases the necessary PV support structures, mainly rooftops and building facades, are there anyway. If a roof or façade is made entirely of PV modules then its cost can be offset against the cost of the building materials it replaces.

Figure 4.10 Proclaiming a message: an eco-home in Denmark (IEA-PVPS).

- BIPV does not require additional land – a very important consideration in urban environments and in countries with high population densities where rural land for PV power plants is expensive and in short supply.
- Renewable electricity is generated and mainly used on site, reducing cable transmission losses.

So how well do photographs included in earlier chapters (Figures 1.13, 1.14, 2.2, 2.23, and 3.14) square up to the expectations of successful BIPV? You may like to refer back and make your own judgements. From the purely technical perspective it is clear that all these PV installations are integrated on to and into the buildings. But from the architectural point of view, overall appearance is key and a PV array should be a harmonious part of the overall design. If these examples underline the difficulty of defining and agreeing architectural aesthetics, this certainly does not absolve us from trying!

4.4.2 PV outside, PV inside

The aesthetics of successful BIPV may be hard to define and judgements are inevitably subjective – yet most of us know instinctively when a solar building feels right for its setting and context. In this section we consider a number of examples to illustrate the wide range of recent international BIPV. Since a picture 'is worth a thousand words' in the field of visual impressions our focus is on photographs accompanied by short explanatory captions.

All PV installations are 'outside' in the sense that they must receive sunlight. Building facades and sloping roofs are often highly visible to the public; flat roofs are more likely to be hidden. Any PV array on public display should appeal to passers-by and bystanders as well as users and owners of a building. Its environmental statement is offered to the world at large.

Although many PV installations are visible only from the outside, some are also 'inside' in the sense that people within the buildings are highly aware of them, and if well designed they can both inspire and delight. Modules may be interspersed with glass windows or arranged as louvres to provide internal shade and ventilation. Some crystalline silicon modules have glass at front and back, allowing light to enter through the gaps between wafers. Thin-film modules can be semitransparent, producing partial shade and generating electricity at the same time. Modules on rooftops that are invisible from the outside may be highly visible on the inside – indeed, this is usually the architect's intention. The advent of tinted and flexible thin-film products means that architects can be increasingly bold and imaginative about incorporating PV into their designs.

It is clear that aesthetic judgements should depend to a considerable extent on whether PV is on the 'outside' or 'inside'. Outside, it interacts with the neighbouring buildings and the local landscape and affects a great many people, some of whom are probably sceptics. Inside, it is more self-contained and speaks only to the users of the building who, in most cases, are enthusiastic supporters of renewable energy. It may be helpful to bear these points in mind when assessing the following photographs. They are arranged in two groups labelled *PV outside* and *PV inside*. The selection is designed to show a good international range of solar buildings with different personalities, acknowledging the efforts that many architects are making to enhance the built environment by incorporating PV imaginatively into their designs.

PV outside

The PV on these buildings and installations is highly visible from the outside.

Figure 4.11 This building in Tübingen, Germany, proudly proclaims its solar identity (EPIA/BP Solar).

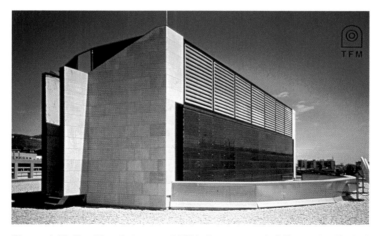

Figure 4.12 Traditional stone and PV in harmony: a building at the Technical University of Catalunya, Spain (EPIA/BP Solar).

Figure 4.13 Architects in countries with a tradition of social housing can spread their influence widely. This example is in Amersfoort, The Netherlands (IEA-PVPS).

Figure 4.14 A Swedish supermarket embraces PV technology (EPIA/NAPS).

Figure 4.15 A huge solar pergola at the World Forum of Culture in Barcelona, Spain, supports a 4000 m^2 PV array (EPIA/Isofoton).

Figure 4.16 The Sydney Olympic Games brought PV to the attention of millions with solar-powered lighting and more than six hundred 1 kW$_p$ arrays on athletes' houses (EPIA/BP Solar).

Figure 4.17 This eco-home in Oxford, England, uses PV modules, water-heating panels and passive solar design to reduce its external energy requirements almost to zero (EPIA/BP Solar).

Figure 4.18 PV louvres replace standard glass shading to provide a dual function (EPIA/BP Solar).

Figure 4.19 A PV-covered walkway at an exhibition centre in Japan (IEA-PVPS).

Figure 4.20 A 1.6 km PV array gives added purpose to a highway sound barrier in Germany (EPIA/Isofoton).

PV inside

The PV on these buildings has a big impact on the internal space.

Figure 4.21 Sunlight and shadow: a striking interior at the Energy Research Centre of The Netherlands (EPIA/ECN).

Figure 4.22 In harmony with nature: 30 kW$_p$ of glass/glass modules at the National Maritime Aquarium, Plymouth, England (IEA-PVPS).

Figure 4.23 Patterned sunlight at the University of East Anglia, England (EPIA/BP Solar).

Figure 4.24 Solar study: the University of East Anglia, England (EPIA/BP Solar).

Figure 4.25 An office interior in Germany (EPIA/Schott Solar).

Figure 4.26 Thin-film semi-transparent modules allow dappled light into this building in Germany (EPIA/Schott Solar).

Figure 4.27 Customer satisfaction: a shop in Tours, France (EPIA/Total Energie).

4.5 Large PV power plants

The current growth, in number and size, of grid-connected power plants is extraordinary.[3] Until quite recently the idea of a PV plant generating megawatts seemed unlikely to most people, but by 2008 there were around 1000 plants worldwide rated at $1\,MW_p$ and above. The great driver of this revolution has been the generous financing of PV electricity in certain countries, most notably Germany, Spain and the USA. Germany and the USA had seen steady increases in capacity for many years; then, in 2007–2008, a remarkable surge took place in Spain due to its government's introduction of a highly attractive tariff of 0.44 euro cents per kWh. In 2008 alone Spain installed $2.7\,GW_p$ of PV, including some $700\,MW_p$ of power plants rated above $10\,MW_p$, in total equivalent to about $50\,W_p$ (a smallish PV module) for every man, woman, and child in the country! When we recall that cumulative global PV production only passed the

Figure 4.28 A Spanish power plant rated at 1.5 MW$_p$. This is close to the average size of large PV plants installed internationally by 2008 (IEA-PVPS).

1 GW$_p$ milestone in 1999 (see Figure 1.11), Spain's achievement in a single year is remarkable. It must be added, however, that the Spanish government reduced the power plant tariff substantially towards the end of 2008, slanting the future more towards roofs and facades, and placed a cap of 500 MW$_p$ on annual PV installation for the following few years. Even though the immediate boom was over, Spain's experience surely changed international perceptions of what is possible, and provided a massive boost to the PV industry.

Other countries active in PV power plant installation are pushing global cumulative capacity into the multi-gigawatt era. Germany and the USA are especially prominent but Japan, Italy, Portugal, France, Greece, and Korea all deserve mention.[3] The international situation in 2008 is summarised by Figure 4.29 for plants above 1 MW$_p$. Part (a) shows that a few of the largest plants already exceeded 40 MW$_p$, but the great majority (830) were in the 1–5 MW$_p$ range. These, together with a large number of lesser installations not shown in the diagram (some on rooftops), produced an overall average size of about 1.25 MW$_p$. About three quarters of plants have static

129

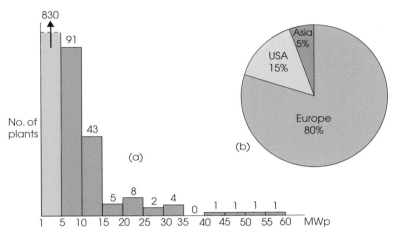

Figure 4.29 (a) The distribution of large PV power plant capacities early in 2008; (b) the contributions to global installed capacity of Europe, the USA, and Asia.

Figure 4.30 The Spanish 9.5 MW$_p$ Milagro Solar Farm with its owners (IEA-PVPS).

arrays; the rest use single- or double-axis tracking, the great majority without concentration. Part (b) of the figure shows the distribution of installed capacity between Europe (mainly Germany and Spain), the USA, and Asia (including a small contribution from the rest of the World).

In such a dynamic situation it is hard to give an accurate snapshot, and even harder to predict what will happen in the coming decade – apart from the

Figure 4.31 La Magascona solar farm in Spain generates 23 MW$_p$ (IEA-PVPS).

near certainty that global power plant capacity will rise dramatically, accompanied by an increase in both peak and average plant sizes. Plants rated at hundreds of MW$_p$ are already on the drawing board and in 2009 it was announced that an installation planned for Mongolia will eventually exceed 1 GW$_p$ in capacity. There is no doubt that large PV power plants have come of age.

References

1. S.R. Wenham *et al. Applied Photovoltaics*, Earthscan: London (2007).
2. F. Antony *et al. Photovoltaics for Professionals*, Earthscan: London (2007).
3. pvresources.com. *Large-scale Photovoltaic Power Plants*, Annual Reports (2007, 2008).

5 Stand-alone PV systems

5.1   Remote and independent

Imagine living in a remote farmhouse, supplied with electricity by an elderly diesel generator and a long way from the nearest electrical grid. The generator needs replacing – but you dislike polluting fumes, the cost of diesel fuel always seems to be rising, and the local electricity utility has just quoted a large sum to connect you to the grid network. How about PV as an alternative? What are the possibilities and pitfalls if you decide on a completely independent stand-alone system?

Figure 5.1 shows a possible scheme. The farmhouse roof faces east–west making it unsuitable for mounting a PV array, so the modules (1) are placed on an adjacent field, south-facing and tilted at an optimum angle. They are interconnected at the array and the DC electricity flows via an underground cable into the farmhouse. The site is windy and exposed so it is decided to include a wind generator (2) in the system. The PV array and wind generator have separate *charge controllers* (3) to regulate the flow of current into a battery bank (4) that acts as an energy store. This is essential because the energy generated by wind and PV is spasmodic and does not coincide with household demand (especially at night in the case of PV!). The battery bank voltage is normally 12 or 24 V DC, but may be higher in a large system. An inverter (5), connected to the battery bank, produces AC at the national supply voltage and frequency (for example 230 V at 50 Hz in Europe, 120 V at 60 Hz in North America and Japan) and supplies the household loads via a fusebox (6), allowing you to use standard AC appliances (7). Note that

Electricity from Sunlight   By Paul A. Lynn
© 2010 John Wiley & Sons, Ltd

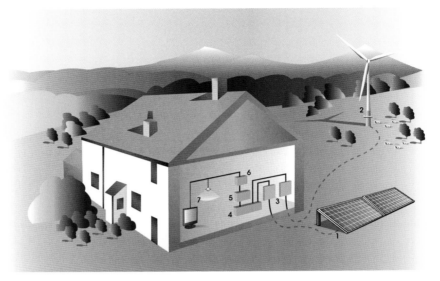

Figure 5.1 Remote and independent: a stand-alone system for a farmhouse.

there is no electric fire in the scheme; generally speaking renewable electricity is too precious to be used for space heating and an alternative such as a wood-burning stove is more suitable.

You may like to contrast this scheme with the grid-connected home illustrated in Figure 4.1. Apart from the wind generator, the major difference between the two systems is the replacement of the grid by a battery bank. Grid connection is relatively straightforward. The PV array in Figure 4.1 is not required to supply all the household's needs, indeed in most cases it supplies considerably less, and the homeowner pays the electricity company for the shortfall. We may think of the grid as an infinite 'source and sink', able to supply or accept any amount of electricity on demand, at any time of day or night. But our stand-alone system enjoys no such luxury. The battery bank is a strictly finite 'source and sink' and its capacity needs careful consideration. Too little capacity, and the electricity supply is unreliable; too much, and the capital cost of batteries becomes excessive. Being autonomous has its problems! As we shall see, the 'sizing' of a PV generator and battery bank to provide an acceptable balance between reliability and cost is a major challenge to the designer of a stand-alone PV system.

As far as the PV modules are concerned, a few points should be added to the account given in Chapter 3. Historically, most PV modules were designed to be suitable for battery charging, and some still are. Typically a crystalline silicon module containing 36 cells connected in series gives an open-circuit voltage of about 20 V and a maximum power point at about 17 V in bright sunlight. This is well suited to direct charging of a 12 V lead-acid battery – the most common type – that reaches about 14.5 V as it approaches full charge. The surplus module voltage is needed to overcome small voltage drops across the blocking diode and charge controller, and to ensure effective charging in reduced sunlight or at high module operating temperatures.

PV modules that are suitable for 12 V battery charging may also be used in grid-connected systems. A good example is the Swiss PV array composed of 36-cell modules previously shown in Figure 3.1. And Figure 2.1 illustrated 72-cell modules that would be suitable for charging a 24 V battery bank. Of course, grid-connection favours higher system voltages with strings of series-connected modules, whereas battery charging requires modules connected in parallel, or series–parallel. In recent years the increasing dominance of grid-connected systems has led manufacturers to offer a wider choice of module sizes and voltages, including many that are not suitable for direct battery charging – a point to be borne in mind when selecting modules for a stand-alone system.

The system shown in Figure 5.1 is fairly sophisticated, involving two sources of renewable energy, battery storage, and an inverter to provide continuous AC power to the household. Various other stand-alone PV schemes are possible, depending on the application. Starting with the simplest, they are:

- *Without battery storage or inverter.* A PV module can supply a DC load directly. A simple example is the type of small solar fountain that floats on a garden pond: the PV sends its current directly to a DC motor driving a pump. The fountain plays only when the sun shines. A more serious application is water pumping for village water supply, irrigation, or livestock watering, where a PV array supplies a DC motor driving a pump that delivers water to a holding tank whenever the sunlight is sufficiently strong.

- *With battery storage, without inverter.* Low-power consumer products such as solar calculators and watches come in this category. So do solar-powered garden lights. Moving up the power scale, a variety of electrical loads, including low-energy lights and a small TV, may be run directly from DC batteries. Many of the solar home systems (SHS's) used in developing countries to supply a small amount of PV electricity to individual families are of this type. A typical SHS comprises a battery, a charge controller and a single PV module (see Figure 1.12). Other examples are DC systems for remote telecoms, security systems, and medical refrigeration.

■ *With inverter, without battery storage.* This type of system produces AC power from a PV module or array and is appropriate when AC electricity is useful at any time of day. For example, AC motors are sometimes used for pumping schemes in preference to DC motors because of their rugged reliability and cheapness (although this must be set against the cost of inverters).

In conclusion, stand-alone PV systems encompass a wide variety of applications with power levels from the miniscule up to a hundred kilowatts or more. Until the 1990s they were the bread-and-butter business of most PV companies, but the recent huge rise in grid-connected systems means that today they account for less than 5% of annual PV module production. However this is not to diminish their huge importance for the families,

Figure 5.2 Off the grid: PV water pumping for a Moroccan village (EPIA/Isofoton).

communities, and businesses that rely on them, including millions of people in developing countries with small solar home systems. In countries such as the USA and Australia there are robust markets for systems in the $1-20\,kW_p$ range, installed on remote farms and in holiday homes that formerly relied on diesel generators (and may still use them for back-up supply). So-called *hybrid systems* integrating PV with wind, hydro, biofuel or diesel generators are attractive where there are large seasonal variations in sunlight levels. And at the top end of the power scale come independent mini-grids in remote mainland areas or on islands, often supplied by several power sources including PV, that provide electricity to whole communities. We shall return to these and other applications later in this chapter. But before venturing out into the wider world we need to discuss batteries, charge controllers and inverters, and explain how a stand-alone system is designed for cost-effectiveness and reliability.

5.2 System components

5.2.1 Batteries

Reliable energy storage is crucial to most stand-alone PV systems. Without it operation of the system is confined to daylight hours when the sunlight is sufficiently strong; with it the user becomes independent of the vagaries of sunlight and can expect electricity by night and day. Many new types of storage battery have come on the market in recent years, including nickel–cadmium, nickel–metal–hydride and lithium-ion, but since the great majority of present stand-alone PV systems use the more traditional lead–acid type we shall concentrate on it in this section.

You are probably familiar with 12 V vehicle batteries, and at first sight they might seem suitable for storing the output of a PV array. But there are important differences between the duty cycle of a standard vehicle battery and a PV storage battery. A vehicle battery's most arduous duty is to supply large currents, typically hundreds of amps, for a very short time to the engine's starter motor. The battery is not supposed to be substantially discharged, except on rare occasions. But a PV battery delivers smaller currents for much longer, and must routinely withstand *cycling*, in other words going through many hundreds or even thousands of charge–discharge cycles without damage. Its duty is rather similar to that of a 'leisure' battery used for running electrical appliances in caravans and boats. But the particular requirements of PV systems in terms of efficiency, reliability, and durability have led manufacturers to develop specialised deep-cycle batteries for the PV market. It should be added that the cheapness and universal availability of standard vehicle batteries means that in practice they are often used in low-power solar home systems (SHSs) in developing countries; and with reasonable success due to the low current levels involved.

High-quality lead–acid batteries for stand-alone PV systems must have long working lives under frequent conditions of charge and discharge. Since PV electricity is precious, especially during long cloudy periods or in winter, the batteries must also display low self-discharge rates and high efficiency. Self-discharge rates of around 3% per month are fairly typical. Efficiency is assessed in three ways:

- *Coulombic* or *charge efficiency*, the percentage of charge put into a battery that may be retrieved from it, typically 85% for lead–acid.

- *Voltage efficiency*, reflecting the fact that the voltage when discharging is less than when charging, typically 90%.

- *Energy efficiency*, the product of coulombic and voltage efficiencies, typically 75%. (Unfortunately some manufacturers quote the coulombic efficiency as 'battery efficiency', which can be misleading).

We see from these figures, and especially the one for energy efficiency, that even high-quality lead–acid batteries cause substantial energy losses in a stand-alone system. Not that all the energy produced by a PV array has to go through the battery charge–discharge process: during periods of strong sunlight the batteries may be fully charged much of the time and the PV electricity can be passed straight to the loads.

So far we have talked about 12 V batteries. But as you are probably aware, a 12 V battery is made up of six electrochemical cells connected in series, each with a nominal voltage of 2 V. A 6 V battery contains three such cells, and so on. High-capacity cells may also be purchased individually and connected in series. Each has a positive and negative electrode made of lead alloy, in an electrolyte of dilute sulphuric acid. Two main categories of cells (and batteries) may be identified:

- *Flooded* or *wet*, using a liquid electrolyte that must be regularly topped up with distilled water. Adequate ventilation must be provided for hydrogen given off during charging.

- *Sealed* or *valve-regulated*, sealed with a gas-tight valve, only allowing gas to escape in the event of overpressure. In normal operation the comparatively small amounts of hydrogen and oxygen produced during charging are recombined to form water, so no topping-up is required. An alternative type of sealed battery uses gel electrolyte. In general, sealed batteries require a strict charging regime, but need very little maintenance.

Batteries recommended for multiple cycling in PV systems often have special electrodes in the form of tubular plates. If not discharged more than 30% they typically survive several thousand charge–discharge cycles; if regularly discharged by 80%, about a thousand cycles.

The capacity of a cell or battery is normally quoted in *ampere hours* (Ah), that is the product of the current supplied and the time for which it flows. For example if a fully charged 12 V battery can provide 20 A for 10 hours, its capacity is 200 Ah (unfortunately many people refer to such a battery as '200 amp'). And since its voltage is 12 V, the total energy stored is $200 \times 12 = 2400$ W h, or 2.4 kWh.

However, it is important to realise that the capacity and energy efficiency of a battery depend on the rate at which it is discharged. The faster the discharge, the lower the capacity. Therefore, when a manufacturer quotes a battery capacity as 200 A h this refers to a particular discharge time such as 10 hours, and this should be specified. The capacity is said to be 200 A h *at the 10-hour rate.* In general we get the most energy from a battery by discharging it as slowly as possible. A 100-hour rate is often considered more relevant to PV applications. Battery capacity also depends significantly on temperature. The rated capacity normally applies to 20 °C and reduces by about 1% for every degree drop in temperature.

We now consider how the voltage of a lead–acid battery varies during charge and discharge. This is very important because, as we shall see in the next section, the charge controller used to regulate current flow from a PV array into a battery (or battery bank) uses voltage as a 'control signal' to protect it from damage and prolong its working life. Once again we can use our 12 V battery as an example.

When a battery is put on charge at constant current its voltage varies in the manner shown in Figure 5.3. Initially close to 12 V, it rises steadily as the *state of charge* (*SOC*) increases. In the final phase it increases more rapidly, reaching over 14 V as full charge (SOC = 100%) is approached. If the battery is of the flooded type this last phase is accompanied by *gassing* in the liquid electrolyte, producing free hydrogen and oxygen. Excessive gassing can occur if charging is continued and may cause damage to the plates; it is very important to provide adequate ventilation to avoid the risk

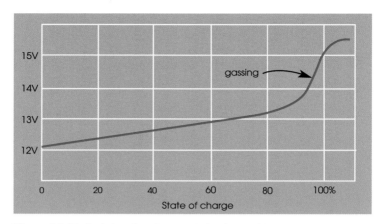

Figure 5.3 Typical charging characteristic of a 12 V lead–acid battery.

of explosion. However occasional, controlled, overcharging known as *equalisation charging* is helpful as the gassing tends to stir up the electrolyte and prevent stratification into different levels of acid concentration. Note that overcharging must always be avoided in sealed batteries, and equalisation is not relevant.

A good charging scheme, which helps keep the battery in top condition is to provide an initial *boost charge* using all the available current; then, as the SOC approaches 100%, an *absorption charge* at constant voltage and low current; and finally a *float charge* to keep the battery gently topped up. Of course, in a PV system dependent on variable sunlight, with none at night, we cannot expect an optimal charging regime. We return to this point a little later.

We next consider what happens during discharge. Figure 5.4 shows typical voltage characteristics when our 12 V, 200 A h battery is discharged at constant current. The curve labelled *10h* is for discharge at 20 A for 10 hours, which reduces the voltage from its starting value down to about 11 V, the point at which the manufacturer recommends disconnection of the load to prevent damage. Note that, at this point, the amount of charge used is 100% – the battery's full nominal capacity. But if we discharge it at the slower rate of 2 A for 100 hours we get the curve labelled *100h*. The voltage holds up better and the total available charge is substantially increased, emphasising once again that the usable capacity of a battery depends significantly on the rate of discharge. It is also very important to note that severe overdischarge of a lead–acid battery, or allowing it to remain at a

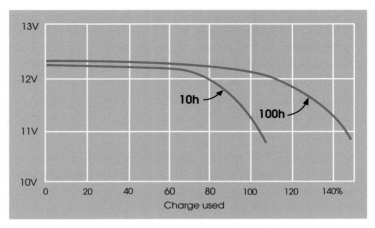

Figure 5.4 Typical discharge characteristics of a 12 V lead–acid battery.

low SOC for lengthy periods, should be avoided whenever possible. In a wet battery the main danger is *sulphation*, the formation of large lead sulphate crystals on the plates, leading to damage and loss of capacity.

In a practical PV system we cannot expect charging and discharging to occur at constant current or in regular cycles of constant depth. The situation is far more complicated and depends on the availability of sunlight compared with the user's demands for electricity. In general we may identify *daily* fluctuations of sunlight and demand, and *seasonal* fluctuations. In sunny summer weather the battery bank is likely to spend more of its time close to full charge (SOC = 100%), with relatively small daily reductions due to demand; but in overcast conditions, or in the winter months, the electricity consumption pattern may lead to periods of low SOC with the risk of supply cut-off. Annual records of charge–discharge cycles in a PV system often appear somewhat random and irregular. Nevertheless, the main points outlined above are useful guides to the performance, care and maintenance of lead–acid batteries, pointing to the ways in which they may be protected by suitable charge control strategies.

5.2.2 Charge controllers

A charge controller is used to regulate the flow of current from the PV array into the battery bank, and from the battery bank to the various electrical loads. It must prevent overcharging when the solar electricity supply exceeds demand, and over-discharging when demand exceeds supply. Various subsidiary control and display functions, depending on the price and sophistication of the unit, are included to protect the batteries from damage and to ensure an operating regime that maximises their performance and length of life. Batteries are an expensive part of most stand-alone systems, especially those required to provide a highly reliable electricity supply day and night, so the relatively modest cost of a good charge controller is money well spent.

In the previous section we saw how the voltage of a battery changes during charging and discharging, and noted that it is used as a control signal to regulate current flow. The two paramount tasks of the charge controller are prevention of battery overcharging, and excess discharging. Overcharging is avoided by disconnecting the PV input whenever the battery voltage reaches an *upper set point*, normally preset at about 14.0 V for float charging, 14.4 V for boost charging, and 14.7 V for equalisation charging of a flooded (wet) 12 V battery. Excess discharge is prevented by disconnecting the load and/or giving a warning by light or sound whenever the voltage falls to a *lower set point*, normally about 11 V. Between these extremes charging and discharging continues in accordance with the amount of sunlight falling on the PV array and the demands of the load.

Ideally, a charge controller continually estimates the battery SOC and uses it to regulate the current accepted from the PV array. Actually, this is more difficult than it sounds because SOC is not simply related to instantaneous battery voltage, but depends on past history. For example if a battery has been supplying load current for some time and its voltage has fallen, then on disconnection it slowly recovers, even without further charging. Conversely if it has been on charge for some time and the voltage has risen, when charging ceases it slowly falls back to a lower level. In other words the voltage signal detected by the charge controller is not a straightforward indicator of SOC. Effective controller algorithms must take past history as well as present voltage into account in assessing SOC and select boost, float, or equalisation charging accordingly.

A closely-related issue is that of *hysteresis*. When the upper set point is reached and the PV array is disconnected to prevent overcharging, the battery voltage immediately starts to fall back, even if no load is connected. How far should it be allowed to fall before reconnection? Too much, and there will be long interruptions to charging; too little, and there will be frequent on/off oscillations. So the gap, or hysteresis, between disconnect and reconnect voltages is a compromise that must be chosen carefully. A similar situation arises at the lower set point. After disconnection at about

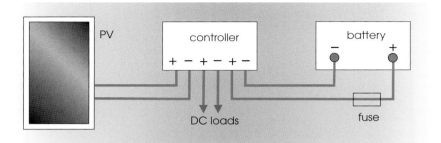

Figure 5.5 A simple scheme for a low-power solar home system (SHS).

11 V, the voltage must be allowed to recover by a reasonable amount before automatic reconnection.

Charge controllers come in many shapes and sizes. In the case of a low-power solar home system (SHS) based on a single PV module and 12 V battery supplying a few low-energy DC lights and a small TV, a simple unit to control a few amps of current at 12 V is all that is required. Figure 5.5 shows the external circuit connections for such a unit. Typically, there is a row of six terminals, one pair each for the PV, DC loads and battery, making installation very straightforward. Note that a fuse has been included close to the positive battery terminal, generally a wise safety precaution in case of a short-circuit.

Moving up the power scale, suppose we have a 1 kW$_p$ PV array feeding a 24 V battery bank with a peak solar current of about 30 A. A suitable controller is likely to offer a number of features such as:

- choice of flooded or sealed lead-acid batteries.
- protection against reverse polarity connection of PV modules or batteries.
- automatic selection between boost, float, and equalisation charging regimes, depending on the estimated state-of-charge (SOC) of the battery bank.
- protection against battery overcharging and deep discharging, excessive load currents, and accidental short-circuits.
- prevention of reverse current at night.
- display of such parameters as battery voltage and/or estimated SOC, PV and load currents, and warning of impending load disconnection.

143

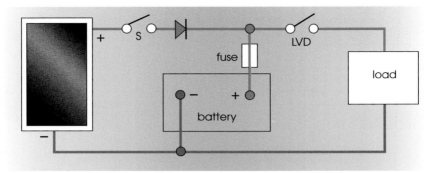

Figure 5.6 Series charge control.

The cost of the unit will clearly depend on how many features are included, and as we move towards the top of the power range, protection and monitoring functions become ever more important and sophisticated.

How do charge controllers perform their central task, regulation of current flow into and out of a battery bank? There are three basic designs on the market: *series* controllers, *shunt* controllers, and *maximum power point tracking* (*MPPT*) controllers. Once again we will illustrate ideas and values using a 12 V system; but they apply equally to higher system voltages if voltage values are scaled up in proportion.

The main functional elements of a series controller are illustrated in Figure 5.6. It includes an electronic switch known as the *low-voltage disconnect* (*LVD*) to prevent battery damage if the voltage falls below some critical value, normally chosen at about 11 V. The diode is included to ensure that reverse current cannot flow back into the PV at night. And to the left of the figure a second electronic switch (*S*), usually a MOSFET, has the vital task of overseeing the charging of the battery. When *S* is closed the PV current is sent to the battery; when *S* is open charging is interrupted. In most modern designs the required switching sequence is achieved by a subtle process known as *pulse-width modulation* (*PWM*). Current is released to the battery in rapid pulses of variable width so that the *average* current, which determines the charging rate, is constantly adapted to take account of the battery's SOC. This is explained by Figure 5.7. The charging rate can be varied continuously between 'OFF' when the battery is fully charged and 'ON' when the available solar current is all sent to the battery. In the 'OFF' condition the pulse width is zero (in effect, no pulses); in the 'ON' condition it is maximum (pulses contiguous). Three intermediate pulse

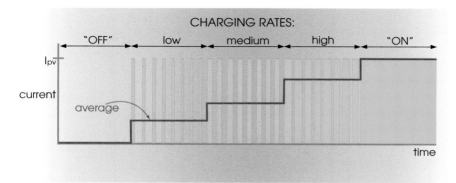

Figure 5.7 Battery charging with pulse-width modulation (PWM).

widths are shown as examples of different charging rates (low, medium, and high). Note that the current switches between zero and I_{PV} the available output from the PV array. The clever part of the PWM approach is, of course, to design a control algorithm that continuously changes the pulse width in sympathy with the SOC, making best use of the PV's output while at the same time protecting the battery.

The operation of a shunt controller is illustrated by Figure 5.8. Here, the electronic switch S is connected across the PV array rather than in series with it, so the battery receives charge when the switch is open. Charging is interrupted when the switch is closed, short-circuiting the PV. Most modern shunt controllers also use PWM to regulate the charging rate. Inclusion of a diode is essential in a shunt controller to prevent the battery from being short-circuited when switch S is closed.

Supporters of the shunt concept often claim that it offers better charging efficiency than the series alternative. Switching losses (which should be small anyway) only occur when the solar current is being rejected; whereas in a series controller switching losses detract from power being sent to the battery. But there are two offsetting disadvantages. First, the shunt controller's switch, normally a MOSFET, needs a larger heat sink and must carry the full short-circuit current of the PV array, possibly in strong sunlight and for long periods. And second, although PV modules do not generally object to being short-circuited, there may be a risk of hot-spot formation due to a 'bad' cell or severe shading (as discussed in Section 3.2.1). In practice there are far more series than shunt controllers on the market, together with a few that combine the two design approaches to produce series/shunt hybrids.

145

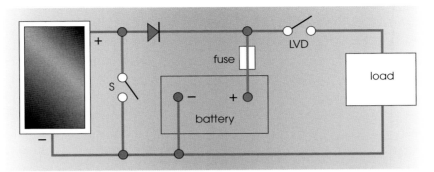

Figure 5.8 Shunt charge control.

We now move to the third basic type of design – the maximum power point tracking (MPPT) controller. Until fairly recently its more complex electronics and higher costs made it something of a niche market product, mainly reserved for the larger stand-alone systems. But as with many aspects of PV, technological advances coupled with volume production have reduced the cost sufficiently to make MPPT controllers attractive in a wide range of systems, even down to a few hundred peak watts. The potential advantage is clear: by working a PV array at its MPP, rather than at a voltage determined by the system's batteries, it is possible to extract considerably more energy and improve system efficiency.

Allowing the PV array to operate at a different voltage from that of the battery bank opens up another important possibility. As we pointed out in Section 3.2.2, the rapid development of grid-connected systems, larger PV modules, and the new thin-film technologies has tended to shift manufacture towards higher module voltages unsuited to the direct charging of batteries. Specifying a MPPT controller allows a wide choice of modules that could not be used with a more conventional series or shunt design.

The basic scheme of a MPPT charge controller is shown in Figure 5.10. The key element is a *DC to DC converter* that allows the PV module or array to operate at a different DC voltage from that of the battery or battery bank. Designs fall into two main categories: *boost converters* that raise the input voltage to a higher level; and *buck converters* that reduce it. Buck converters are more common in PV applications, reducing the relatively high voltages of modern PV modules (or series-connected modules) to the lower voltages of battery banks. DC to DC converters have undergone extensive development in recent years and their ability to 'transform DC'

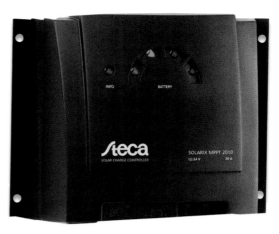

Figure 5.9 This MPPT controller can control a 12 or 24 V system with PV array power up to 500 W$_p$ and MPP voltages up to 100 V. With dimensions 19 × 15 × 7 cm, it weighs 900 g (Steca Elektronik GmbH).

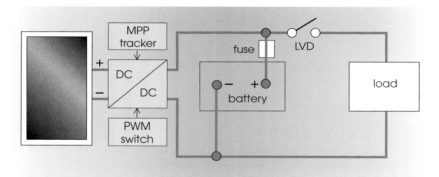

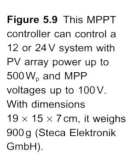

Figure 5.10 Extracting the most from a PV array: the MPPT charge controller.

finds many applications in electronic engineering. In the case of a MPPT charge controller, the really innovative part is not the voltage change, but rather the ability to sense the MPP of the PV array as sunlight levels change, at different times of day and in variable weather. Typically this is achieved with an algorithm that performs continuous electronic tracking of the array's MPP, together with periodic sweeps along its *I/V* characteristic to confirm that the true MPP is being detected rather than a local power maximum. The voltage on the input side of the converter is automatically adjusted to the MPP voltage.

One further feature is required, this time on the output side of the DC to DC converter. As in more conventional charge controllers of the series or shunt type, the available output from the PV array, at its new voltage level, is presented to the batteries as a train of rapid current pulses of variable width using a PWM switch. Once again, an advanced control algorithm ensures that charging rate is continuously adapted to the estimated SOC of the batteries.

5.2.3 Inverters

We described a stand-alone PV system for a remote farmhouse at the start of this chapter (see Figure 5.1). It includes an inverter, connected to the battery bank, for supplying AC to the various household electrical appliances. Of course, inverters are not required in systems having only DC loads; but when they are used, it is important to understand the special features required in independent, stand-alone, units. As the stand-alone PV market develops customers are increasingly opting for AC systems because they like the flexibility offered by a wide range of consumer electronics, household appliances, electric tools, and even washing machines. AC systems are also used by hospitals and remote telecommunications sites, and for running machinery in small factories. Well over 50% of newly installed stand-alone systems are AC.

The grid-connected inverters described in Section 4.2 are not suitable for stand-alone systems. An important difference is that whereas a grid-connected inverter must generate AC at precisely the right frequency and phase to match the grid supply, a stand-alone unit is not so constrained. It generates its own AC without any need to lock into a grid and is necessarily self-commutated. Although the generated waveform must suit the various AC loads it need not satisfy an electric utility. And whereas a grid-connected inverter is supplied directly from a PV array and often performs MPP tracking, its stand-alone cousin is fed from storage batteries at a more or less constant DC voltage, leaving the task of extracting energy from the PV array to a charge controller that may itself work on the MPPT principle.

Figure 5.11 shows the connections for a typical mid-range stand-alone system with a PV power between (say) 1 and $2\,kW_p$. It is similar in many ways to the much lower-power, DC only, solar home system in Figure 5.5, with an inverter added; however it is redrawn to emphasise several features of a larger, more sophisticated, system:

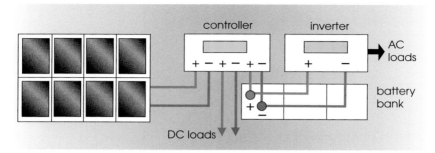

Figure 5.11 Typical connections for a mid-range stand-alone system.

- The PV is an array rather than a single module.
- A battery bank replaces a single battery, giving more storage capacity.
- The charge controller has an electronic display (or a set of coloured LEDs) indicating parameters such as battery voltage, SOC, PV current and load current.
- The inverter, connected directly to the battery bank, also indicates its operating conditions.

We have not shown fuses because adequate fusing is normally included in the charge controller and inverter. Note also that many controllers supply 12 V loads; if these are not required, the system voltage may be set to a higher value such as 24 or 48 V. Some controllers offer dual-voltage operation, typically 12 or 24 V, but limited to the same maximum current. So they can regulate a more powerful system, provided the higher voltage is selected. Since the inverter is connected directly to the battery bank it should include a disconnect function if the battery voltage falls below the lower set point.

In Section 4.2 we considered ways in which the choice between various types of inverter is influenced by module connection schemes and the problem of unavoidable shading. Although similar considerations apply in principle to the stand-alone case, in practice the situation tends to be simpler. Shading is rarely a problem because PV arrays may often be sited on open ground, with all modules facing the same direction, rather than confined to potentially awkward urban rooftops; this, together with the

149

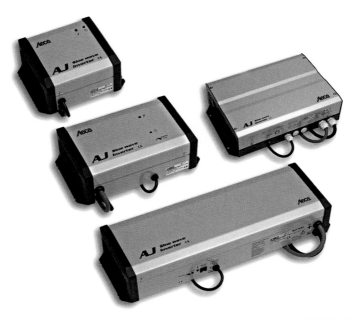

Figure 5.12 This family of inverters covers the power range 200 W to 2 kW (continuous), with system voltages of 12, 24, and 48 V (Steca Elektronik GmbH).

moderate size of most stand-alone systems, means that a single central inverter is generally suitable.

The user of a stand-alone inverter should look for the following technical features:

- A power rating sufficient for all loads that may be connected simultaneously.

- Accurate control of output voltage and frequency, with a waveform close to sinusoidal (low harmonic distortion), making the AC supply suitable for a wide range of appliances designed to run off a conventional electricity grid.

- High efficiency at low loads, and low standby power draw (possibly with automatic shut-down when all loads are turned off), to avoid unnecessary drain on batteries.

- Ability to absorb or supply reactive power in the case of reactive loads.

■ Tolerance of short-term overloads, particularly caused by motor start-up.

Inverter efficiency is especially important in a stand-alone system that must obtain all its energy from precious sunlight without grid back-up. Maximising efficiency and minimising standby consumption do not come cheap, but the resulting energy savings may allow the system designer to specify a smaller PV array and battery bank, leading to overall cost savings. Unfortunately some inverter manufacturers only quote maximum efficiency, or efficiency at full rated output, disguising unfavourable performance under low-load conditions. Figure 5.13(a) shows two typical efficiency curves, red for an inverter incorporating a low-frequency transformer, orange for a unit with a high-frequency transformer. Both suffer from severely reduced efficiency when delivering less than about 10% of their rated output. With rising load the efficiency reaches a maximum over 90% and then tails off again. But there are subtle differences between the two: the unit with the low-frequency transformer does better at low load, and vice versa (switching losses associated with HF electronics are relatively dominant at low load, whereas magnetic losses in a low-frequency transformer are greater at high load). Such differences can be important when choosing an inverter for a particular duty.

Many stand-alone systems, including those in solar homes, spend much of their time on low load with peaks at certain times of day. Figure 5.13(b) shows a representative daily load profile for a home running a wide range of electrical appliances for lighting, cooking, and household machines. Most of the time the inverter load is less than 20% of its rated output, with peak periods in the morning and evening. This is exactly the sort of situation where careful attention to the inverter's low-load efficiency and standby power requirements is likely to pay dividends.

Like other aspects of PV engineering, the stand-alone inverter scene is advancing rapidly. AC systems are finding favour for small solar home systems in developing countries, and individual PV modules with integrated inverters have entered the market. Some manufacturers offer inverters combined with charge controllers in single units; others use a modular design approach so that many inverters can be stacked together to increase power-handling capacity. Towards the top end of the power range there is ever-increasing sophistication in monitoring, data logging, and intelligent power management. And some of the most powerful units are being used as central inverters for mini-grids of $100\,kW_p$ or more, often integrating PV with other energy sources and providing renewable electricity to remote villages and island communities.

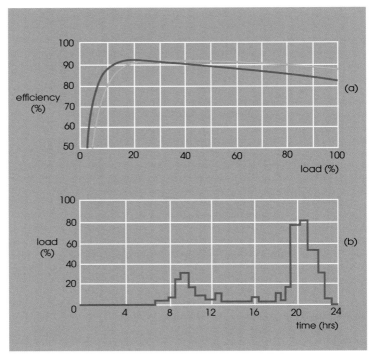

Figure 5.13 (a) Efficiency curves for two types of inverter; (b) a daily load profile for a solar home.

5.3 Hybrid systems

Stand-alone systems that rely on natural energy flows in the environment must inevitably cope with intermittency. Their main defence against unreliability and loss of service is a battery bank to store incoming energy whenever it is generated and feed it out to the electrical loads on demand. But in many cases system reliability may be enhanced, and the size of the battery bank reduced, by a hybrid system based on two or more energy sources. PV and wind power are often attractively complementary, especially in climatic regions such as Western Europe where low levels of winter sunshine tend to coincide with the windiest season of the year (and, of course, wind does not refuse to blow at night!). You may have seen examples of small PV–wind hybrid systems at roadsides, powering traffic control or telecommunications equipment. We illustrated a larger one for a remote farmhouse in Figure 5.1. Worldwide, many large hybrid systems are based on the valuable partnership between PV and wind.

Stand-alone electrical systems in isolated areas, including those for homes and farmsteads, are often referred to as *remote area power supplies* (*RAPS*), a market traditionally satisfied by diesel generators. For those of us who like to champion renewable energy it may be rather hard to extol the virtues of hybrid systems based on PV and diesel fuel, but they do offer advantages in many practical situations and are widely used. The benefits may be summarised as follows:

- It may too expensive, in terms of the PV array and battery store, to provide a sufficiently reliable service with photovoltaics, especially where solar insolation is highly seasonal. For example, does it make economic sense to install a PV system that can cope with occasional high load demands in winter when sunlight is in short supply? A hybrid system with a back-up diesel generator may be a better option.

- Diesel engines are very inefficient when lightly loaded, giving poor fuel economy. Low running temperatures and incomplete combustion tend to produce carbon deposits on cylinder walls (glazing), reducing service lifetimes. It is advisable to run engines above 70–80% of full rated output whenever possible. But a lone diesel generator that can cope with occasional peak demands is likely to run at low output much of the time. Better to turn it off and use PV and the battery bank when electricity demand is low. The diesel can boost charge the batteries if necessary, at a high charging rate.

- In addition to rising fuel costs, unpleasant fumes, and the noise of diesel engines, it may be difficult to obtain reliable fuel supplies and engine maintenance services in remote locations. PV needs no fuel and, provided the battery bank is looked after properly, should be low-maintenance.

- If an existing diesel installation needs upgrading, the addition of PV may be a good solution. Being essentially modular, PV may be added in small stages, raising system power capacity in line with increasing demand.

We see that the combination of PV with diesel can offer distinct environmental and economic benefits compared with a diesel generator on its own. Each energy source is used to best advantage, taking account of its special

153

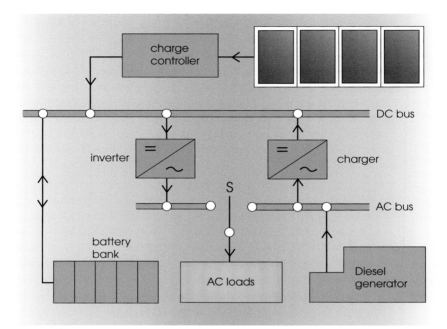

Figure 5.14 A PV–diesel hybrid system.

features. Substantial savings on diesel fuel and maintenance can be realised in those hybrid systems where a diesel generator remains the most realistic option for meeting occasional high load demands and providing security of supply.

Figure 5.14 illustrates a common form of PV–diesel hybrid system. The PV array feeds its electricity into a main cable highway known as the *DC bus* (short for busbar) via a charge controller, and the diesel generator supplies AC electricity to an equivalent *AC bus* that supplies the AC loads. The AC bus and DC bus are connected by an inverter and a battery charger, which may be combined in a single unit. This allows the diesel generator to charge up the battery bank if required; and the battery bank to supply AC to the electrical loads. A master switch *S*, operated either manually or automatically, effects changeover between the diesel generator and the battery bank for supplying the AC loads, depending on operating conditions. Intelligent use of this arrangement ensures that the diesel engine is always run fairly hard to satisfy a high load demand or boost charge the battery bank. At other times the PV and battery bank take over.

This system, in which the AC loads are switched between the diesel generator and the battery bank plus inverter, is conceptually simple and quite common in practice. It is straightforward to implement as a system upgrade for an existing diesel installation. An alternative *parallel-hybrid* configuration dispenses with the changeover switch and uses automatic control circuits and a more sophisticated inverter-charger to bring in the diesel generator when necessary.[1] Such a system can often meet the load demand in a more optimal way without the need for human supervision. A fuller account of the technical, economic, and environmental aspects of diesel hybrid stand-alone systems is given elsewhere,[2] and we will meet a sophisticated island mini-grid of this type in Section 5.5.2.

We conclude this section by a return to our starting point: the potential of hybrid systems, including those based on PV and wind energy, to raise the reliability and reduce overall costs of renewable electricity in remote areas. In principle it is possible to include several different sources (not necessarily including diesel generators). Care is needed over system integration, for example in choosing several stand-alone inverters which cannot, in general, be interconnected because of the need to synchronise their AC outputs in frequency and phase. But, once again, modern electronics including power-conditioning and control units come to the rescue, with increasingly elegant solutions to the needs of the PV systems engineer.

5.4 System sizing

5.4.1 Assessing the problem

In the popular imagination science provides firm answers to firm questions, leaving little to chance when it comes to technical decision-making. But things are not as simple as that. For example, while almost all experts agree that global warming due to greenhouse gas emissions poses a major threat to life on Earth, there are wide-ranging views on its exact severity and timescale because the supporting evidence is essentially statistical. Scientists and engineers are trained to understand technical uncertainty, but it often confuses the public, and offers scope for vested interests to declare the whole idea erroneous or exaggerated.

In this book most of our discussion is based on 'hard science' and we have been able to describe the performance of individual system components such as PV modules, batteries, and inverters with considerable accuracy. But there are two major chapters in the PV story where chance and uncertainty play a key role. Interestingly, but perhaps unsurprisingly, they are at opposite ends of nature's range of operations – one dealing with the miniature, the other with the large-scale. The miniature, discussed in Section 2.2.3.3, concerns the quantum nature of light and the random way in which solar photons are absorbed or transmitted by solar cell materials. Although we avoided the mathematical details, you may be sure that the underlying theory is replete with probabilities! The second topic, the large-scale one we are about to tackle, concerns system sizing – deciding how much PV power and battery storage is needed for a particular stand-alone system, based on estimates of local insolation patterns, electricity demand, and the required reliability of service. A few moments reflection will surely convince you that such estimates must always be hedged about with uncertainty. Indeed, so much so that the 'sizing problem' is often considered the most difficult aspect of system design.

This is primarily a stand-alone rather than a grid-connected problem because independent systems lack the support of a powerful electricity grid acting as a flexible 'source and sink'. A stand-alone system, especially when powered by PV alone, cannot realistically achieve total reliability. There is inevitably a trade-off between reliability and cost, forcing the system designer (and customer) to face some difficult choices. We can illustrate the dilemma using four stand-alone PV scenarios with very different operational expectations:

- *PV in Space.* Launched into Space on long missions without any prospect of replacement or repair, the PV arrays on spacecraft are surely the most extreme examples of stand-alone systems. 100% reliability is certainly the aim, probably over many years and at almost any cost because spacecraft are entirely dependent on their PV power supplies. Fortunately, there is one simplifying factor: insolation in Space, beyond the Earth's volatile atmosphere, is highly predictable, removing one major source of design uncertainty.

- *PV-powered refrigeration.* PV is increasingly used to power refrigerators for storing vaccines and medicines in remote hospitals in developing countries. Failure of the electrical supply may be life-threatening as well as highly inconvenient and expensive, so reliability is obviously a major requirement.

- *PV-powered traffic signs.* Also a 'professional' application, traffic signals to warn drivers that they are speeding, or that there is an obstruction ahead, should obviously be dependable. But how dependable, and over what timescale? What if the PV electricity runs out for a few days, and foggy weather makes an accident more likely? Will the highway authority's budget stretch to units containing more PV and larger batteries?

- *PV for a solar home or farmhouse.* We have already illustrated a stand-alone system for a farmhouse (see Figure 5.1). The size of the PV array and battery bank will obviously depend on the input from the wind turbine, the owner's choice of electrical appliances, and the amount of use. There is plenty of room for flexibility and cost-saving here although it may be very difficult to decide such issues at the design stage. Generally speaking, security of supply is judged less important than for the 'professional' systems mentioned above, even though to be without lights and a TV in dead of winter is not an attractive option! In a holiday home used mainly in the summer months, occasional supply failure may be quite tolerable.

It is clear from the above examples that the designer of a stand-alone PV system is faced with difficult decisions and choices. They can be approached in various ways. Sizing methods based on practical experience and 'rules of thumb' are quite often used and may provide sensible, cost effective solutions without much appreciation of the background science. PV sizing software is also widely available, although there is always a danger of using inaccurate input data or failing to appreciate the underlying assumptions. At the other extreme, analytic methods that attempt to put figures (including probabilities) to the many individual factors and components in a PV system promise more accuracy and scientific insight, yet they are also highly dependant on the robustness of input data and assumptions.[3] Our own approach, similar to one recommended elsewhere,[4] is intermediate in sophistication yet sufficiently detailed to highlight the main technical issues. We will illustrate it with an example based on a holiday home in southern Germany that is mainly used in the summer months.

We start the design process by considering the range of electrical appliances required by the homeowners, the power that each appliance consumes, and the average amount of daily use. This allows us to specify the total amount of electricity required in an average day, which is basic information needed to size the PV system. The table shown in Figure 5.15 includes eight low-energy lights and a TV (often considered priorities for homeowners) plus

Appliance		Power (W)	No.	Average hrs/day	Average Wh/day
Light		11	8	3	260
TV		60	1	4	240
Computer		60	1	3	180
Refrigerator		80	1	24 (on-off)	500
Kettle		1000	1	0.2	200
Microwave Oven		700	1	0.4	280
Food Mixer		400	1	0.15	60
Washing Machine		800	1	0.6	480
				Total	2200

Figure 5.15 Appliances and energy requirements for a stand-alone system.

a number of other appliances reflecting individual needs and preferences. Note that by multiplying the power of each by its estimated average daily use, we arrive at its consumption in watt hours (Wh) per day and, at the bottom of the table, the total estimated consumption for the whole system – in this case 2200 Wh (2.2 kWh) per day. This is the amount of electrical energy to be supplied by the PV system and is fairly typical for a solar home system (SHS) that includes a good range of modern appliances (by contrast, simple SHSs in developing countries based on a single PV module and a battery often provide just 200–300 Wh/day). In this case the homeowners wish to use standard AC appliances, so an inverter must be included in the system.

It is extremely important to specify the most energy-efficient appliances available and, wherever possible, to avoid those involving heating. Electric fires for space heating must be considered taboo, PV electricity being far too precious to be used for warming human bodies! The kettle in the above list might also be thought extravagant; a daily consumption of 200 Wh is sufficient to provide about ten cups of coffee or tea and whether its great convenience is worth the energy cost is clearly a personal choice. The same applies to the microwave cooker (but note that it is only switched on for very short periods). A washing machine is high on many people's list, but it must not be used to heat the water; running it once or twice a week rather than daily would be very helpful. In short, everything should be done to reduce daily usage, especially of high-consumption units, with the aim of reducing the size and cost of the PV system. We are here confronting a reality that escapes most people living in developed economies: electricity cannot always be taken for granted and used casually, but must sometimes be treated as a precious resource.

Having decided on the daily amount of electricity required, we are ready to tackle two key aspects of system design – the power of the holiday home's PV array, and the capacity of its battery bank.

5.4.2   PV arrays and battery banks

In the previous section we estimated 2.2 kWh as the average daily electricity requirement for a holiday home in southern Germany, and it is now time to decide on the amount of PV and battery storage needed to meet the specification. In this section we shall often refer to arrays and battery banks, terms appropriate for medium-size and larger systems, but our approach is also valid in principle for small systems containing a single PV module and battery.

The first task is to work out the size of the PV array: how much peak power should it have to satisfy the electricity demand? As it stands, the 2.2 kWh/day applies throughout the year whereas the amount of sunlight falling on the array is bound to be seasonal. So if we size the array to cope with the 'worst' month for sunlight – usually December in the northern hemisphere – the owner is likely to be paying a lot of money for an array that is unnecessarily powerful in summer. Since this is a holiday home, it may be more sensible to restrict the 2.2 kWh daily usage to the summer months.

We will assume that the array can be sited on adjacent open ground, facing south and inclined at a suitable tilt angle. Back in Section 3.3.2 we discussed the amount of daily solar radiation falling on south-facing inclined PV arrays and Figure 3.13 showed typical data in the form of monthly mean values for London and the Sahara Desert. We also introduced the concept of *peak sun hours* for estimating an array's annual output. This involves compressing the total radiation (direct plus diffuse) received throughout the year into an equivalent duration of standard 'bright sunshine' ($1\,kW/m^2$). The same concept may be used for daily radiation. For example, if an inclined array receives an average insolation of $3\,kWh/m^2$ per day in April, this is considered equivalent to 3 peak sun hours; so an array rated at (say) $2\,kW_p$ is predicted to yield $3 \times 2 = 6\,kWh/day$. Although it is an approximation that tends to be over-optimistic for arrays receiving a high proportion of diffuse radiation, it offers a very straightforward way to estimate array output in a particular location.

Figure 5.16 shows daily solar radiation levels in the same form as Figure 3.13, but using published data[5] relevant to the holiday home's location in southern Germany. Three representative values of tilt are illustrated: $33\,°$; $48\,°$ (the latitude angle); and $63\,°$. As expected, $33\,°$ does best over the summer months when the Sun is high in the sky (an even smaller tilt would give better results at midsummer, but at the expense of other times of year). A tilt of $48\,°$ gives good results around the time of the equinoxes in March and September; and $63\,°$ is marginally preferable over the winter months. We also see that radiation levels in December are only about one third of those in midsummer, so a PV array big enough to supply the home's electricity in December would be three times oversized in June. Clearly,

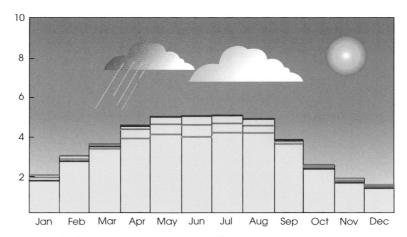

Figure 5.16 Daily solar radiation in kWh/m^2 on south-facing inclined PV arrays for a location at latitude 48 °N in southern Germany. Three values of array tilt are illustrated: 33 ° (blue); 48 ° (red); and 63 ° (green).

choosing an array to cope with the 'worst' month of the year would be a very expensive option.

At this stage the system designer must surely discuss alternatives with the homeowners. For example they might agree to restrict their demand for 2.2 kWh/day to the months March to September, covering the main holiday period, in return for a smaller PV system at lower cost. Over this 7-month period the 33° tilt angle is a good choice. The 'worst' month is now taken as March, for which the average daily radiation is 3.5 kWh/m². This figure can be used for sizing the array. The homeowners will have to make do with considerably less electricity over the winter months, unless the total is boosted by an alternative energy source. Or perhaps they will agree to forgo use of the refrigerator, microwave oven and washing machine, and cut down on the drinking of coffee! Unlike the 'professional' PV systems mentioned in the previous section, a 'leisure' installation should offer plenty of opportunities for energy saving, trading convenience and reliability against cost.

Using the peak sun hours concept we may express the average daily amount of electricity available for running the home's appliances, E_D as:

$$E_D = P_{PV} S_p \eta \tag{5.1}$$

Where P_{PV} is the rated peak power of the PV array, S_p is the number of peak sun hours per day in the month of interest, and η is the overall system efficiency (discussed below). Therefore the peak power of the array is given by:

$$P_{PV} = E_D / S_p \eta \tag{5.2}$$

In the case of the holiday home, E_D = 2.2 kWh/day, S_p = 3.5 h in March, and we will assume a system efficiency of 60% (η = 0.6), so that:

$$P_{PV} = 2.2/(3.5 \times 0.6) = 1.05 \text{ kW}_p \tag{5.3}$$

We therefore predict that a PV array rated at just over 1 kW$_p$ will supply the daily load requirement of 2.2 kWh during the months March to September.

The overall system efficiency η takes account of various power reductions and losses that prevent the PV array's nominal output from getting through to the household's AC appliances. A figure of 60% may seem disappointing, but is fairly typical of such stand-alone systems. It is derived by multiplying together efficiencies for the various system components, expressed as numbers between 0 and 1 (for example, an efficiency of 85% is expressed as 0.85). Although it is difficult to give exact figures the following are fairly typical:

- *PV modules (0.85).* Power output is less than the rated value in standard 'bright sunshine' ($1\,kW/m^2$), due to such factors as raised cell operating temperatures, dust or dirt on the modules, and ageing. Also, modules are not generally operated at or close to their maximum power point (unless a controller with MPP tracking is used).

- *Battery bank (0.85).* The charge retrieved from the battery bank is substantially less than that put into it (see Section 5.2.1).

- *Charge controller, blocking diodes, and cables (0.92).* There are small losses in all these items.

- *Inverter (0.9).* This is a typical figure for a high-quality inverter, bearing in mind that it must sometimes work at low output power levels (see Section 5.2.3).

The product of all these figures is 0.6, or 60%. If MPP tracking is used and the system is DC only (no inverter), the system efficiency might approach 70%. But in practice it is hard to predict how components will behave in variable sunlight and ambient temperatures, or how the system will actually be used by the homeowners as they become familiar with it, so the above figures should be treated with caution.

In view of all these uncertainties, plus the vagaries of the weather, oversizing a PV array by a reasonable amount – say 20% – is often recommended. In the above example $1.2\,kW_p$ would obviously improve reliability of supply, but it is, as ever, a question of cost. An alternative is to regard PV as an essentially modular technology that can easily be upgraded. So it would be possible to install a $1\,kW_p$ array initially, and expand it later if required.

The remaining task is to size the battery bank. The biggest decision is how many 'days' of battery storage are required. Too few, and a spell of unusually dull or wet weather may cause a serious loss of electricity supply. Too many, and the battery bank becomes unnecessarily large and expensive. Five days of usable battery storage (in the above example, equal to $5 \times 2.2\,kWh = 11\,kWh$) is often regarded as a good compromise between reliability and cost. But of course it depends on the application; a holiday home is by no means a crucial case and many 'professional' systems demand far higher reliability to avoid risking serious inconvenience, economic penalties, or even danger to life. In such cases the amount of battery storage may have to be raised greatly, perhaps to 15 days or more. Alternatively, a reliable standby power source such as a diesel generator may be incorporated.

When the number of days of storage N has been decided, the capacity C of the battery bank can be calculated:

$$C = N\,E_D/D\,\eta_{inv} \tag{5.4}$$

Where (as before) E_D is the daily electricity requirement, D is the allowable depth-of-discharge of the battery bank, and η_{inv} is the efficiency of the inverter, assuming an AC supply is required. Note that the usable capacity of the battery bank is less than its nominal, rated, capacity because complete discharge must be avoided. In our example we will assume 5 days of storage, battery discharge up to 80% of nominal capacity, and an inverter efficiency of 90%. Hence:

$$C = 5 \times 2.2/0.8 \times 0.9 = 15.3 \text{ kWh} \tag{5.5}$$

As with the PV array, it may be sensible to oversize the battery bank some-what – or to treat it as modular, with the option of upgrading it later.

To summarise, the stand-alone system for the holiday home in southern Germany should be able to supply the desired amount of electricity between March and September using a PV array rated at $1.05\,kW_p$ with a battery bank of capacity 15.3 kWh (assessed at the 100-hour discharge rate normal for PV systems). If the batteries are connected to give 24 V DC, which is quite common for a system of this size, then the required charge capacity is 15300/24 = 638 A h.

This specification could be met, with a reasonable amount of oversizing, by an array of (say) eight PV modules rated at $150\,W_p$ each ($1.2\,kW_p$ total), together with a bank of (say) eight 12 V batteries rated at 175 A h each (16.8 kWh total). The electricity yield, and hence system reliability, could be further improved at modest cost by specifying a MPPT charge controller. The modules could be connected in series, or series–parallel; the batteries as four parallel strings of two units each to give 24 V DC. The main components of the system are illustrated in Figure 5.17.

We end this section with some further remarks on reliability. First, it must be admitted that choosing a holiday home to illustrate system sizing makes life rather easy because it allows a somewhat cavalier approach towards possible supply failures. We are making use of the relative unimportance of failures in this 'leisure' application and have assumed that homeowners are flexible over their use of appliances. All this changes dramatically in the case of a 'professional' system with stringent load and reliability criteria, and serious thought must given to how often a failure of supply can be tolerated.

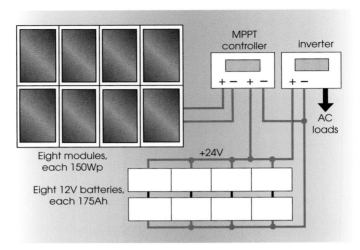

Figure 5.17 A suitable system for the holiday home.

The many uncertainties of system design mean that the problem can only be discussed sensibly in terms of probabilities. A measure known as *loss-of-load probability* (*LLP*) is widely used. Basically, LLP denotes the probability that, at any point in time, the PV system is unable to satisfy the demand for electricity. It may also be interpreted as the proportion of total time that the system is unavailable (which should include estimated maintenance and repair outages). LLP = 0 implies that the system is 100% available; LLP = 1 that it is permanently out of action. We normally hope for and expect low LLP values, say between 0.0001 and 0.1, but it depends very much on the importance of reliability in a particular application. The smallest values of LLP, increasingly difficult and expensive to achieve, are typically found in PV systems used on space missions or in vital telecommunications links, the largest ones in leisure applications (it is hardly a disaster if solar-powered garden lights fail to work every evening!).

However it is much simpler to explain the LLP concept than to calculate its value for a particular system, or design a new system to meet a customer's LLP specification. The basic difficulty is that it depends on so many factors, some fairly obvious, others obscure or random in nature. Our previous discussion makes clear, for example, that reliability is generally increased (and LLP reduced) by specifying a more powerful PV array and/ or a larger battery bank – although there is, in fact, a subtle interaction between them.[2] Sunlight statistics play a major role. For example, occasional lengthy periods of cloudy weather, untypical of the local climate, can result in a battery bank's state-of-charge (SOC) being depleted to such an extent that supply cut-off is inevitable. Unfortunately rare and isolated weather events cannot be predicted from averaged meteorological data.

Yet in spite of the difficulties, various theoretical ways of incorporating LLP into stand-alone system design have been developed in the past 25 years,[2] and many sophisticated computer programs for system sizing and simulation are available.[6] Indeed, the complexity of the task more or less demands the use of computer software, even though it may be hard for the newcomer to understand its details. A straightforward quantitative approach to sizing, such as we have introduced in this section, seems a good antidote to over-reliance on computer software. A few simple calculations at least allow us to check that the numbers churned out by a computer program are reasonable!

5.5 Applications

The variety of applications for stand-alone PV systems is extraordinary. Almost any need for electricity in isolated, remote, or independent locations can, in principle, be met by solar cells. We have already mentioned a number of examples in this book, from solar-powered watches and calculators to space vehicles, but our main focus has been on electricity supply for remote buildings far from an electricity grid. This has provided a chance to describe typical units that make up medium-power systems, including PV arrays, battery banks, charge controllers, and inverters, in a setting that most of us can easily imagine. It is now time to move out into the wider world – and beyond – to discuss a number of key application areas where PV has made, and continues to make, major contributions. It is hard to select just a few examples from the large number of possibilities; so we have chosen four distinctive topics, each important in its own way, that illustrate a wide range of issues and challenges in PV system design.

5.5.1 PV in Space

For more than half a century spacecraft have relied on solar cells for their power supplies. In the early years of the modern PV age solar electricity was so expensive that space exploration provided its only significant market. The costs of designing, manufacturing, and launching vehicles into Space are so large that the price of cells to power them is relatively

Figure 5.18 PV encircles the Earth (NASA).

unimportant, the main criteria being technical performance and reliability in the harsh space environment. Although the total amount of PV power launched beyond the Earth's atmosphere is tiny compared with today's gigawatts of terrestrial installations, solar cells remain vital to modern spacecraft including those used for satellite communications, TV broadcasting, weather forecasting and mapping – and, of course, space exploration. You may wish to follow up this brief introduction with an approachable and authoritative account given elsewhere.[7]

We may summarise the special features of the space environment that impact on the design and deployment of solar cells and arrays with a few key points:

- Radiation in Space tends to damage solar cells.
- Sunlight in Space, unfiltered by the Earth's atmosphere, has a different spectrum from that received by terrestrial PV cells.
- Spacecraft, including satellites in earth orbit, experience dramatic changes in sunlight intensity and temperature as they move in and out of shadow, causing high thermal stresses in solar cells and modules.

- PV modules and arrays need to be kept as small, neat and light as possible to avoid adding unnecessarily to the launch payload.
- Sustained technical performance and reliability are paramount, especially on long missions.

Each of these will now be discussed in more detail.

Radiation damage to solar cells in Space is a major challenge to PV designers. The risk of damage by high-energy electrons and protons is particularly serious for satellites in *mid-Earth orbits* (*MEOs*), defined as 2000–12 000 km above the Earth, which pass through the *Van Allen* radiation belts. The neighbourhood of Jupiter is also a high-radiation environment. Special types of cover glass are effective at reducing the steady and cumulative degradation of cell performance over the lifetimes of long missions. The susceptibility of standard silicon solar cells to radiation damage was recognised in the early years of space exploration and much effort has been put into design improvements to mitigate the effects and raise cell conversion efficiencies, presently approaching 20%. Although high-efficiency silicon cells are in widespread use, a major advance in recent years has been the development of multi-junction III–V cells based

Figure 5.19 Wide horizons for PV: the Solar System (NASA).

on gallium arsenide and related compounds, which are much less suscep-tible to radiation damage and offer even better conversion efficiencies. We first described these cells in Section 2.4.3.1, and since they are so important to space PV we will mention them again towards the end of this section.

In this book we have often referred to standard 'strong sunlight' received by solar cells and modules at the Earth's surface. This is defined as having an intensity of $1 \, kW/m^2$ and the AM1.5 spectrum typical of sunlight after passing through the Earth's atmosphere. Sunlight in Space, unfiltered by the atmosphere, is described by the Air Mass Zero (AM0) spectrum (both spectra were illustrated in Figure 1.6). The intensity also varies according to distance from the Sun. For example near Mercury it is almost double that near Earth, near Jupiter only about one-thirtieth. Clearly, solar cells and modules have to operate satisfactorily and be calibrated for use in such conditions.

Closely related to changes in light intensity are changes in operating tem-peratures. Whereas terrestrial solar cells and modules are normally required to work between, say, $-20 \, °C$ and $+70 \, °C$, conditions in Space can be far more demanding. Spacecraft in orbit around the planets experience extremes of temperature as they pass in and out of the Sun's illumination. Cell tem-peratures in *low earth orbit* (*LEO*) may get down to $-80 \, °C$ in shadow, but in orbit around distant Jupiter they must work at $-125 \, °C$, even when illu-minated; around Mars, at up to $+140 \, °C$. Sudden transits from shadow to sunlight can produce big power surges as well as exposing cells and modules to high thermal stresses.

The size and weight of space PV arrays is extremely important, as is their ability to deploy successfully on reaching zero-gravity conditions. The smaller and neater an array, the easier it is to integrate into the spacecraft's structure during launch; the lighter it is, the lower the payload and launch cost. For a given peak power, an array's area is proportional to cell effi-ciency, favouring the most efficient cells and technologies. In the early 1970s the most powerful PV system in Space was that of the *Skylab 1* satel-lite, delivering about $16 \, kW_p$. The *International Space Station*, launched in 1998 and continuously expanded and developed over the following decade, generates more than $100 \, kW$ of average power from its silicon solar cells, which are mounted in eight double-arrays with a total area of over $3000 \, m^2$. This is large-scale PV, with exciting technical challenges for project teams in mechanical engineering and materials science as well as solar technology.

The efficiency of solar cells designed for use in Space is important for several interrelated reasons. For a given peak power requirement, improve-

Figure 5.20 The International Space Station, photographed in 2009 (NASA).

ments in cell efficiency reduce the area, weight, and payload costs of a PV array. As we mentioned earlier, one of the most important advances in recent years has been the commercial development of triple-junction cells based on gallium arsenide (GaAs) and related compounds, which now attain 30% efficiency under AM0 conditions, reducing array areas by over a third compared with high-efficiency silicon. They also have better radiation resistance. Typically, a triple-junction device consists of a 'sandwich' of layers of gallium indium phosphide (GaInP), gallium arsenide, and germanium (Ge), each carefully chosen to absorb a portion of the solar spectrum – you may like to refer back to Section 2.4.3.1 and Figure 2.33 for a fuller explanation. Research continues apace, with even more efficient four-junction devices in prospect, and an increasing interest in concentrator systems to reduce the area and cost of these highly specialised cells.

It goes without saying that technical performance and reliability, sustained over long periods, are paramount in space systems. On manned missions there may be limited potential for carrying out maintenance and repair; but on long unmanned missions solar cells and arrays are quite literally on their own – surely the most extreme example of stand-alone systems. It is hardly surprising that PV power systems in Space cost hundreds of times more per peak watt than their earthbound counterparts; but without them space-craft would, quite literally, be lost.

5.5.2 Island electricity

Providing a small island community with an economic, convenient, and reliable electricity supply can be a major challenge. Traditionally, islanders in the developed world have installed diesel generators and depended on fuel deliveries from a mainland depot. But diesel engine maintenance is expensive, fuel costs always seem to be rising – and there is a noise and pollution problem that people who cherish their natural environment would rather avoid. Most islands have a valuable wind resource, many have lots of sunshine and free-running rivers or streams. Such plentiful flows of natural energy act as a strong incentive to generate renewable electricity, and when several different energy sources are available it makes good sense to consider a hybrid system and distribute the electricity using an island mini-grid.

Such systems are still 'stand-alone' in the sense of being unsupported by large conventional electricity grids. So are mini-grids serving isolated communities on the mainland. Their major advantage compared with individual stand-alone systems for each user is that integration of various energy sources with different daily and seasonal peaks can provide a more consistent, reliable and economic supply for a whole community. Although back-up diesel generators are generally still needed to ensure a reliable 24-hour service throughout the year, they can be started up for short periods only when necessary – and then run hard and at high efficiency.

The Isle of Eigg, 6×4 km in extent, is one of the jewels of the Inner Hebrides. Lying off the west coast of Scotland to the south of Skye, it has an equable climate thanks to the Gulf Stream, a generous wind resource, lots of sunlight in summer, a few streams, and just under a hundred inhabitants. Like many Scottish islands, Eigg has a harsh history behind it, including 19th-century depopulation and more recent absentee landlords, but in 1997 funds were raised to purchase the island and set up the Isle of Eigg Heritage Trust to manage it for the inhabitants and their wonderful environment. Determined to update their electricity supply from reliance on ageing diesel generators to a modern 'green' alternative, they raised capital grants totalling £1.6 m for a hybrid system comprising PV, wind, and hydroelectric power, with diesel back-up.[8,9] Early in 2008 all 37 households and 5 businesses on Eigg were connected to the new island grid, achieving celebrity status for a state-of-art renewable energy system that is

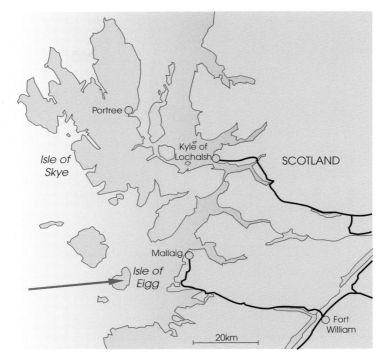

Figure 5.21 The Isle of Eigg lies off the west coast of Scotland.

providing inspiration to other island communities in Scotland and around the world.

Eigg's system is illustrated in Figure 5.22, which summarises the generation and consumption of electricity. A key feature is that all generators and loads are interconnected by an island-wide AC grid. Transmission is at 11 kV for long cable runs and at 230 V for short ones (from the PV and diesel generators), with transformers inserted where necessary. Power sources that generate DC (the wind turbines and PV) feed into the grid via inverters. An advanced load management system monitors the balance between supply and demand, bringing in the diesel generators when necessary, and controlling energy flow to and from the battery banks via a set of bidirectional inverter–chargers. The batteries, PV and wind turbines are the only DC components; homes and businesses are supplied with 230 V AC. Grid frequency is set by the inverter-chargers, or by the diesel generators when they are running. We now comment further on the various items:

171

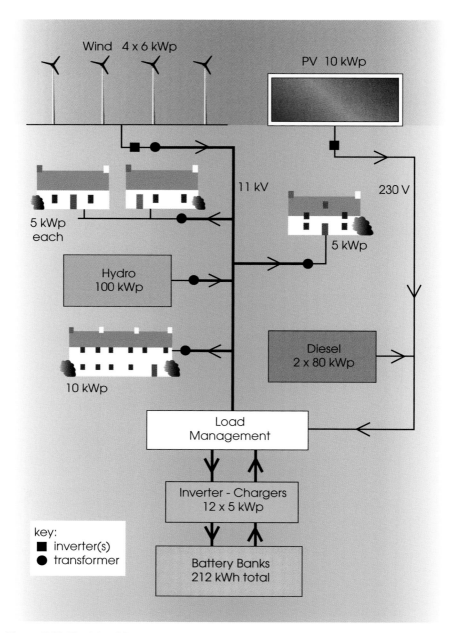

Figure 5.22 The Isle of Eigg's renewable energy system.

■ *10 kW$_p$ of PV.* It may be surprising to see a substantial PV array included because Scotland is hardly noted for its sunshine! However the Hebridean islands have a better sunshine record than the mainland, where higher mountains tend to increase cloud formation and precipitation. Eigg, at latitude 57 °N, has plenty of sunlight in the summer months, with up to 18 hours between sunrise and sunset in June, so PV can make a valuable contribution when wind and hydropower tend to be at their lowest. In this system the output from 60 PV modules, connected in 6 strings of 10, is converted to 230 V AC by adjacent inverters.

■ *24 kW$_p$ of wind energy.* A group of four wind turbines, each rated at 6 kW$_p$, is sited at one of the island's windiest locations. Although wind turbines are generally rated in kilowatts at a standard high wind speed, we have used kW$_p$ units in the figure to emphasise that they rarely operate at full output – even though the months October to April are highly productive on an island subject to Atlantic storms. In fact the Eigg wind turbines make a valuable contribution throughout the year. Their DC outputs are inverted and transformed to 11 kV for transmission.

■ *100 kW$_p$ of Hydropower.* The most powerful contributor to the renewable energy portfolio is a new run-of-river water turbine rated at 100 kW$_p$ supplied by a substantial stream (there are also two much smaller pre-existing turbines in other locations rated at 9 and 10 kW$_p$, not shown in the figure). However on a small island the flow of streams closely follows current rainfall and tends to be intermittent

(a)   (b)

Figure 5.23 PV on Eigg (Wind & Sun Ltd); windpower on Eigg (Eigg Electric).

and seasonal. Hydroelectric generation on Eigg is therefore variable, much stronger in winter than summer, with an average value far lower than the nominal ratings of the turbines.

■ *2 × 80 kW$_p$ of diesel.* Two new diesel generators provide back-up to ensure 24-hour service throughout the year. In an average year the renewable sources are expected to provide over 95% of total electricity demand, so the total diesel contribution is small. Typically, the generators are run hard for short periods to boost-charge the battery bank on days when the renewables are unable to meet the full load demand. They generate power at 230 V AC.

■ *Load management.* A comprehensive hybrid system of this kind, involving various energy sources and domestic and business loads, justifies a sophisticated control system. Its aim is to make the most of available renewable generation, deciding between the various sources in times of surfeit, ensuring that the battery bank is neither overcharged nor over-discharged, transmitting electricity efficiently to the various loads, and bringing in the diesel generators when necessary.

■ *12 × 5 kW$_p$ inverter–chargers.* Arranged as four 3-phase clusters, each with its own battery bank, the bidirectional inverter-chargers are at the heart of the system. When the renewable generation is insufficient to meet demand they take energy from the batteries and invert it to augment the AC supply. When generation exceeds demand they rectify the AC and charge the batteries. If the batteries are fully charged and excess energy is being generated, the inverters raise the frequency, and additional 'opportunity' loads such as heaters in community buildings (not shown in the figure) detect the increase and switch on automatically. If there is still surplus energy, the frequency is increased further and the various generators respond by backing off to prevent battery overcharging.

■ *4 × 53 kWh battery bank.* The batteries are arranged in four 48 V banks and located in the power house with the inverter–chargers and diesel generators. The banks are normally kept above 50% state-of-charge (SOC) to prolong their life. The quoted total capacity is therefore half the full nominal capacity of 424 kWh and equates to approximately one day's electricity usage on the island. Additional days of storage are not needed in this case because of the diesel back-up.

■ *Households and businesses.* 37 households and 5 businesses were initially connected to the island grid and supplied at 230 V AC.

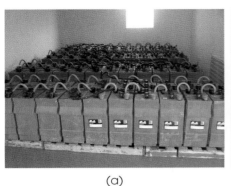

(a)   (b)

Figure 5.24 The battery banks, and some of the main inverter-chargers (Wind & Sun Ltd).

Householders agreed to limit peak demand to $5\,kW_p$ each, businesses to $10\,kW_p$. All consumers are provided with an energy meter to monitor the amount of electricity being used. The islanders have adapted well to the new system and are far better informed about electricity usage and energy conservation than most people on the mainland.

Although the Eigg electricity supply is not especially strong in PV, it is an excellent example of a modern hybrid system. The PV component, being essentially modular, may be increased in the future to provide more summer electricity. In any case, the principles of design and implementation are of widespread relevance, even though the relative contributions from PV, wind, hydro and diesel back-up power are bound to vary from one island system to another.

5.5.3 PV water pumping

Infectious diseases caused by tainted drinking water and primitive sewage disposal are largely unknown to those of us who live in the developed world. We tend to take the benefits of pure water for granted. But whom should we thank for this blessing? It has been said that the civil engineers of the nineteenth century did more to improve public health than all the doctors and surgeons put together, by designing and building the infrastructure for modern water supplies.

The situation can be very different elsewhere. In rural areas of some of the poorest countries in the world millions of people, especially women, spend

Figure 5.25 Clean and accessible: PV-pumped water (EPIA/Schott Solar).

hours each day fetching and carrying water, sometimes from polluted streams or pools. Yet new village wells can transform lives and health and, if equipped with automatic pumps, eliminate the daily grind of water collection.

Water pumping is one of the most successful applications of stand-alone PV in developing countries. By the year 2000 over 20 000 PV-powered systems were in use worldwide and the pace of installation continues. Of course, small water pumps can be worked by hand, larger ones by windmills or diesel engines. But the PV alternative, in addition to its cleanliness, reliability and long life, often proves economic for medium-size systems. Water pumping is also used for crop irrigation and stock watering.[2]

A typical scheme for village water supply is illustrated in Figure 5.26. A submersible pump/motor, protected by installation underground, raises water to a storage tank whenever sunlight falling on the PV array is sufficiently strong. From there it is fed by gravity to one or more taps. In previous sections we have often discussed the need for battery storage in stand-alone systems. But one major feature of water pumping is that the water tank replaces batteries as the energy store, using PV electricity directly to increase the potential energy of the raised water.

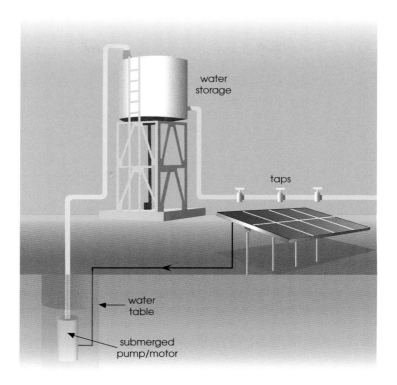

Figure 5.26 A system for village water supply.

Although the scheme is simple in principle, a number of technical choices must be made:

- *Type of pump.* Of the many types of pump on the market, *centrifugal* designs are widely used to raise water against pumping heads up to about 25 m (the height difference between the water table and tank's input pipe). Multi-stage versions can cope with higher heads. A centrifugal pump has an impeller that throws water against its outer casing at high speed, the kinetic energy then being converted to a pressure head by an expanding output pipe. Centrifugal pumps are compact, robust, and well-suited to PV applications, but they are not normally self-priming and must therefore be kept submerged. This makes them suitable for pump/motors positioned below the water table. Alternative *displacement* or *volumetric* pumps including various self-priming types are more suitable for lower flow rates from very deep wells or boreholes.

■ *Type of motor.* DC motors are generally more efficient than AC ones, but more expensive. AC motors are very rugged and need little or no maintenance, so are suitable for submersion at the bottom of a well; but inverters are needed to convert PV electricity to AC, adding to the capital cost. Among DC motors the *permanent-magnet* type is often preferred; but all conventional designs use carbon brushes that must be periodically adjusted or replaced, making submersion awkward. Modern *brushless* DC motors overcome this difficulty, at a cost.

■ *Matching the motor and PV array.* Ideally, the PV array should be operated close to its maximum power point (MPP) in all sunlight conditions. Unfortunately the resistive load offered by most motors does not allow this to happen, so a MPP tracking controller based on a DC to DC converter may be inserted to improve matching and increase efficiency.

From the PV perspective the most important task is to size the array to pump the desired amount of water. A well-known hydraulic equation is a good starting point. The hydraulic energy E_h required to raise $1\,m^3$ of water against a head of H_w metres is given by:

$$E_h = \rho g H_w \quad \text{joules} \tag{5.6}$$

where ρ is the density of water ($1000\,kg/m^3$) and g is the acceleration due to gravity ($9.81\,m/s^2$). In this case H_w is the height of the holding tank's input pipe above the water table.

In this book we have generally used the kilowatt hour (kWh) as our unit of energy. We note that 1 joule is equivalent to 1 watt second, or $1/3.6 \times 10^6\,kWh$. Therefore if we wish to raise a volume V_w cubic metres of water per day, the required hydraulic energy is:

$$E_h = V_w\,H_w\,(1000 \times 9.81/3.6 \times 10^6) = 0.0027\,V_w\,H_w \quad kWh/day \tag{5.7}$$

For example, suppose that a village population of 300 needs an average of 50 litres of water per person per day – a total of 15 000 litres or $15\,m^3/day$ – and that the tank's inlet pipe is 20 m above the water table. The hydraulic energy required is:

$$E_h = 15 \times 20 \times 0.0027 = 0.81\,kWh/day \tag{5.8}$$

Figure 5.27 PV for a village water supply in Niger (EPIA/Photowatt).

We can now estimate the size of the PV array using the peak sun hours concept first mentioned in Section 3.3.2. Basically, this involves compressing the total daily radiation received by the array into an equivalent number of hours of standard 'bright sunshine' ($1\,kW/m^2$). The peak power of the array is then approximately given by:

$$P_{PV} = E_h/S_p\,\eta \tag{5.9}$$

where S_p is the number of peak sun hours for the particular location and η is the overall system efficiency. The peak sun hours are normally chosen for the 'worst' month to ensure continuity of supply throughout the year. The system efficiency must take account of electrical losses in the motor and cabling, hydraulic and friction losses in the pump and pipework, and mismatch between the motor and the PV that prevents the array from working at its maximum power point. An average efficiency of about 40% is fairly typical for a centrifugal pump and, together with other losses, gives a typical system efficiency of around 25% (0.25). So, as an example, if the location has an insolation equivalent to 3 peak sun hours per day in the 'worst' month, then the PV array needs a peak power:

179

Figure 5.28 A large PV water-pumping station in Morocco (EPIA/Isofoton).

$$P_{PV} = 0.81/3 \times 0.25 = 1.08 \text{ kW}_p \qquad (5.10)$$

Although peak powers up to a few kilowatts are fairly typical of systems supplying water to individual villages in sunshine countries, considerably larger PV arrays are sometimes installed to serve larger communities – for example, a cluster of villages obtaining water from a single source that is distributed by hand or pipe. A good example is the Ouarzazate scheme in Morocco, consisting of more than 20 autonomous stand-alone systems supplying a total population in excess of 10 000 people. One of the smaller PV systems in this scheme has already been shown in Figure 5.2; a much larger one, incorporating a substantial roof-mounted PV array, is shown in Figure 5.28 – an impressive example of PV water pumping in action.

5.5.4 Solar-powered boats

Boats powered by sunlight represent one of the most successful and attractive applications of PV in the field of sustainable transport. Less well-known to the public than the solar car races that have achieved international fame in Australia and the USA, solar boating has recently made headlines with a growing number of international events and a transatlantic crossing. Solar circumnavigation of the globe is a definite prospect. Unlike road vehicles, boats do not have to climb hills or travel at high speed and they require surprisingly little power for propulsion in calm conditions. This makes solar-powered boating on lakes, rivers, and canals relatively inexpensive and opens up a new market for PV in an important leisure industry.

The low power levels needed to propel boats at modest speeds in calm water can be nicely illustrated with a historical example. Two hundred years ago Britain was in the middle of a canal-building frenzy. The heavy materials of the early industrial revolution, including coal and iron, needed to be transported over considerable distances for which the road network was totally inadequate. So the English narrow canals, with locks just over 2 m wide and 22 m long, were carved through the countryside by gangs of 'navvies' (derived from the word navigation) using picks, shovels, wheelbarrows and human muscle power. This extraordinary feat of civil engineering revolutionised inland transport and allowed cargos up to about 30 tonnes to be carried in individual barges, the so-called narrowboats, that just squeezed into the locks. And how did the boats move in those early days? They were towed, often two at a time, by a single horse! Admittedly at slow speed, typically 2–3 km/h (kph), but it was a vast improvement on existing methods of transport by land.

This example suggests that a single 'horsepower', nowadays taken as equivalent to 746 W, is enough to shift many tonnes of boat at modest but useful speeds. And if careful attention is paid to design by making hull, motor, and propeller as efficient as possible, we now know that one or two horsepower (HP) can propel a modern leisure craft with several passengers at realistic speeds – say up to 10 kph (6.4 mph) in calm water. The quest for efficiency mirrors that of solar car design with its emphasis on streamlined bodywork and high-performance motors, transmission, and tyres. But in the case of boats the power levels, and therefore costs, tend to be much lower.

Electric boats are a novelty to many people. For the last hundred years most motorboats have used petrol or diesel engines for propulsion, helping to deplete the Earth's valuable fossil fuels, making a lot of noise and polluting the waterways. But it was not always so. In the period from the 1880s up to the start of the First World War in 1914 there were plenty of battery-powered electric boats on the lakes and rivers of Europe, including some that could carry over 50 passengers. The river Thames in England boasted a scheduled passenger service, with electric charging stations along the bank. However the advent of internal combustion engines proved nearly fatal and by 1930 electric boating was in severe decline. Half a century later it began to emerge again, largely due to increasing environmental awareness, and today represents a small, but flourishing sector of the leisure boating industry. The essential components – batteries, control circuits, electric motors and propellers – are constantly being developed and refined, giving wonderfully silent cruising with minimal disturbance to wildlife and riverbank.

Solar electric boats are even more of a novelty. We are not talking about the many boats that use a PV panel or two to power their electronic equipment and cabin lights, but true electric boats that use PV for propulsion. These exciting craft literally 'cruise on sunlight'. Today there are many examples on the inland waterways of Europe, North America, and Australia, and the number rises year by year. The combination of a virtually silent, nonpolluting electric drive and solar energy is extremely attractive.

As already noted, quite a lot can be achieved with a propulsive power of 1 HP, equivalent to 746 W. In fact the range 200 W to 3 kW covers most modern electric leisure boats at normal cruising speed, and there are a few larger craft, including passenger ferries, that require considerably more. We are referring to the mechanical power needed to propel the boat forward; more electrical power is required because of combined motor and propeller losses, typically amounting to 40%.

We now describe three recent boats with different design criteria, specifications, and passenger accommodation. The first, 6.2 m catamaran *Solar Flair III*, cruises on inland waterways in England. Designed as an experimental boat to test various combinations of PV modules, motors, and propellers, she also appears at boat shows and rallies, helping to promote PV and solar boating and convince the public of its viability, even in the British climate. She carries six $75\,W_p$ monocrystalline silicon PV modules in front of a small cabin, plus two more behind (not visible in the photo), giving a total of $600\,W_p$ to charge batteries that power an electric outboard motor. A smaller additional motor, mounted below the front module, acts as a bow thruster to aid sharp turning on narrow canals and rivers. The main motor takes about 450 W of input power to attain a cruising speed of 8 kph in calm conditions. Average summer sunshine produces enough PV electricity to move *Solar Flair III* about 32 km (20 miles) per day at this speed. The design aims at technical performance and a streamlined appearance rather than passenger accommodation.

Figure 5.29 Solar-powered catamaran *Solar Flair III* (Paul A. Lynn).

Our second example, the 6.7 m (22 ft) pontoon boat *Loon*, has been designed and developed in Ontario, Canada, as a spacious canal and river cruiser able to accommodate up to eight passengers in comfort. Raising the $1\,kW_p$ of PV modules on a canopy greatly increases passenger space and gives protection against rain – and maybe also sun! The input motor power to achieve 8 kph is about 1 kW and the PV provides enough electricity, in the Canadian summer months, to travel an average of about 24 km (15 miles) per day at this speed. On long cruises the boat's batteries may be fully recharged by plugging into shore power electricity.

The third example, 14 m Swiss catamaran *Sun21*, made history in 2007 by completing the first Atlantic crossing entirely on solar power. She carries $10\,kW_p$ of crystalline silicon PV modules on a canopy, and needs about 3.8 kW of motor power to cruise at 8 kph in calm water. On the Atlantic voyage the PV provided up to about 45 kWh/day and since the boat was travelling day and night the motor input power had to be kept down to an average of around 1.5 kW, giving a speed of about 5 kph (3 knots) in sea conditions. *Sun21* is an impressive catamaran with accommodation for five crew members. Before the Atlantic voyage very few people believed that

Figure 5.30 The pontoon boat *Loon* (Tamarack Lake Electric Boat Company).

a motorboat, travelling entirely on sunlight, could achieve such a feat and she received a rapturous welcome on reaching New York.

The catamaran or pontoon form of hull is very popular for solar-powered boats, with sleek twin floats providing a good stable platform for PV, especially when raised on a canopy. However there is nothing to stop designers from using conventional monohulls; the main criterion is an efficient low-drag hull that creates minimal wash and uses the precious PV energy to best advantage.

Finally, we consider the question 'What exactly makes a boat solar-powered?' Exaggerated claims are sometimes made; it is easy to stick a PV module or two on a boat, and claim that it is powered by the sun. But it does PV no good to overstate its performance and capabilities, leading to disappointment and scepticism. One answer is to use a simple measure known as the *solar boat index* (*SBI*) to quantify performance and allow sensible comparison of a wide variety of boats carrying different amounts of PV.[10]

Figure 5.31 *Sun 21* arrives in New York (Dylan Cross Photographer).

The SBI is based on the peak sun hours concept introduced in Section 3.3.2. We have also used it to size PV arrays for water pumping in the previous section. It involves compressing the daily radiation received by an array into an equivalent number of hours of standard 'bright sunshine' ($1\,\text{kW/m}^2$). In this case the most relevant radiation data is that for a horizontal surface (most PV modules on boats are mounted horizontally) during the summer months of the boating season. An array rated at peak power P_{PV} watts and receiving an average S_p peak sun hours per day is expected to yield about $S_p\,P_{PV}$ watt hours per day. If the boat needs an input motor power P_M watts to cruise at a standard speed (normally taken as 8 kph) in calm conditions, then the SBI is defined as:

$$SBI = \eta\,S_p\,P_{PV}/P_M \tag{5.11}$$

where η is a system efficiency that accounts for the PV generally operating away from its maximum power point (MPP), and for battery storage losses. Using typical figures of 80% (0.8) for the PV and 75% (0.75) for the batteries, the system efficiency $\eta = 0.8 \times 0.75 = 0.6$. If we now assume $S_p = 5$

(typical daily peak sun hours for midsummer in Western Europe), Equation (5.11) becomes:

$$SBI = 3\,P_{PV}/P_M \tag{5.12}$$

This is easy to remember and is in fact used in the UK to quantify the performance of solar-powered boats.[10]

The SBI has a simple interpretation. It represents the approximate number of hours per day, in average summer weather, that a boat can travel at standard speed on its PV electricity. For example if a boat's SBI is unity, this means it can travel about 1 hour a day, or 7 hours a week at 8 kph to give a range of 56 km. Most inland leisure boats are weekend boats, for which this amount of cruising is fairly typical. Therefore it seems reasonable to describe leisure boats with SBI values of 1.0 or above as 'solar-powered' in the West European and similar climates; otherwise they are 'solar-assisted'. Although the SBI is only approximate, it does provide a simple quantitative measure of a boat's cruising range on sunlight, and allows the solar performance of different boats to be compared. The SBIs for the three examples are:

> *Solar Flair III:* 4.0   *Loon:* 3.0   *Sun21:* 7.9

Clearly, these values need sensible interpretation because the patterns of use of the three boats are different and so are the solar climates in which they operate. What we can say is that, if the three boats met together on a European lake, their SBIs should give a good indication of relative solar performance.

Worldwide, there are a number of competitions for solar-powered boats that act as good catalysts for new ideas and designs, encouraging young people to get involved. A good example is the Frisian Solar Challenge,[11] held biannually on canals and lakes in the Netherlands. Such events do an excellent job of bringing to public attention the exciting future of solar-powered boats with their silence, lack of pollution, and minimal environmental impact.

5.5.5 Far and wide

The applications described in previous sections represent a broad range of technical, economic, and social objectives. Yet the scope and geographical spread of stand-alone PV systems stretch much wider. We end this chapter with a few more photographs and captions to illustrate some of PVs past and present successes and help stir the imagination for its future potential.

On land and sea

Figure 5.32 Two solar-powered cars, entered by the Universities of Michigan and Minnesota, speed at over 100 kph along a Canadian highway during the 2005 North American Solar Challenge (Wikipedia).

Figure 5.33 It has become commonplace for sailors to install PV modules on the decks of ocean-going yachts to power cabin lighting, services, and navigation equipment. There is now growing interest in making the sails themselves 'photovoltaic' (EPIA/Shell Solar).

In heat and cold

Figure 5.34 This installation in the Libyan desert provides *cathodic protection*, an important application of PV that helps minimise corrosion of metal structures including pipelines (EPIA/Shell Solar).

Figure 5.35 A PV array produces electricity for a meteorological station in Greenland. In this high northern latitude the vertical array captures much of the available sunlight, and solar cell efficiency is enhanced by the very low temperatures (EPIA/Shell Solar).

For education and information

PV school in China
© IT Power

Figure 5.36 An increasing number of schools worldwide use PV arrays to generate valuable electricity and stir their students' imagination for the future of renewable energy. But it is unusual to find a large stand-alone system like this one in China (EPIA/IT Power).

Shell Solar B.V.

Figure 5.37 Another example of a large PV array in a remote location: this one helps to transmit information by telecommunications link (EPIA/Shell Solar).

189

References

1. T. Markvart (ed.). *Solar Electricity* (2nd edition), John Wiley & Sons, Ltd: Chichester (2000).
2. S.R. Wenham *et al. Applied Photovoltaics*, Earthscan: London (2007).
3. A. Luque and S. Hegedus (eds.). *Handbook of Photovoltaic Science and Engineering*, John Wiley & Sons, Ltd: Chichester (2003).
4. F. Antony *et al. Photovoltaics for Professionals*, Earthscan: London (2007).
5. NASA (eosweb.larc.nasa.gov/sse), *Surface Meteorology and Solar Energy Tables* (2010).
6. S. Silvestre. Review of System Design and Sizing Tools, in T. Markvart and L. Castaner (eds). *Practical Handbook of Photovoltaics*, Elsevier (2003).
7. S. Bailey and R. Raffaelle. *Space Solar Cells and Arrays*, in reference 3 above.
8. Eigg Electric (www.isleofeigg.net/trust/eigg_electric.htm). *Isle of Eigg Electrification Project* (2010).
9. Synergie Scotland (www.synergiescotland.co.uk). *Eigg Electrification Project*, case study 20 (2010).
10. P.A. Lynn. What is a Solar Boat? *Electric Boat News*, **18**(4) 13 (2005). See also www.electric-boats.org.uk
11. Frisian Solar Challenge (www.frisiansolarchallenge.nl). *Frisian Solar Challenge: World Cup for Solar Powered Boats* (2010).

6 Economics and the environment

6.1 Paying for PV

6.1.1 Costs and markets

One of the most encouraging aspects of the current PV scene is the steady reduction in costs. Continuing improvements in cell and module efficiencies are making a substantial contribution; but above all it is the sheer volume of production in state-of-the-art factories using highly automated facilities that is driving down costs. Right back in Section 1.4 we introduced the 'learning curve' concept to illustrate how, for a wide range of manufactured products, costs tend to fall consistently as cumulative production rises. Figure 1.11 confirmed that PV costs have fallen for more than two decades by around 20% for every doubling of cumulative production – and the trend continues. The long-held, almost cherished, ambition of the PV community to produce modules at 'one US dollar per watt' was finally achieved in 2009 in the case of high-volume thin-film CdTe manufacturing, with rival cell technologies not far behind.

Of course the cost of a PV system also depends heavily on balance-of-system (BOS) components and there are design, installation, and maintenance charges to consider. Fortunately, most of these are also falling broadly in line with cumulative PV production and today typically represent – as they have in the past – about half of total system costs.

Electricity from Sunlight   By Paul A. Lynn
© 2010 John Wiley & Sons, Ltd

The speed of market penetration by a new technology normally depends greatly on economics. Potential purchasers of grid-connected PV systems, which have come to dominate the global market, wish to know how much solar electricity costs to generate. For example, if you are considering installing a rooftop PV system, how does the cost of a unit of electricity (1 kWh) compare with the price charged by the local utility, and does it look like an attractive investment? In the case of stand-alone PV systems there are different criteria since grid electricity is not generally available as an alternative; comparisons are more likely to be made with diesel generators, and decisions affected by environmental concerns, including noise and pollution.

It is important to bear in mind that, in many cases, the installation of a PV system is not only about money. Companies may be concerned to demonstrate their green credentials, schools to educate and inspire their pupils, and individuals to 'do their bit' to reduce carbon emissions. You may know someone who, instead of buying an expensive new vehicle, settled for a cheaper model that burns less fuel and spent the rest of the money on a rooftop PV system. For citizens in developed economies it can be as much a lifestyle choice as a purely economic one.

As far as the economic case is concerned, Figure 6.1, although necessarily speculative, illustrates some important trends. Predicted costs of PV elec-

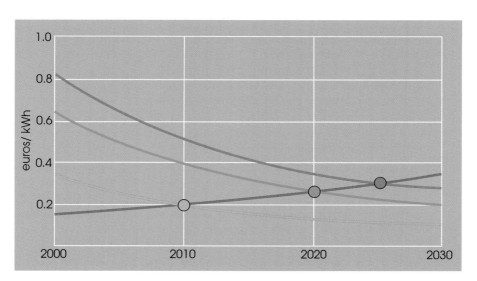

Figure 6.1 Towards grid parity in Europe.

tricity in euros/kWh are plotted up to year 2030 for electricity supplied by utilities to domestic customers in Europe (red curve); and for electricity generated by rooftop grid-connected PV systems in various countries (orange, green and blue curves). Most experts expect that the increasing global demand for energy, together with falling fossil fuel reserves, will result in real price rises for conventional electricity in the coming years. This is shown by the red curve, assuming an annual increase of 2.5% compound. By contrast, the price of solar electricity is expected to fall as cumulative PV production soars. In sun-drenched European locations such as southern Spain and Italy (orange curve), the current cost is roughly competitive with conventional electricity because PV arrays are highly productive. In less sunny northern Germany and England (green curve), PV is expected to achieve 'grid parity' by about 2020; in Norway and Sweden (blue curve), perhaps 5 years later. But whatever the detailed timescales, the trends seem clear and inevitable – even if the citizens of northern Europe will need a bit more patience!

In many ways this picture is oversimplified. First, the costs of PV systems and the prices paid by consumers for grid electricity are not uniform between different countries. Second, price increases for grid electricity over the coming years cannot be predicted with any certainty. And additional factors will surely influence the cost of PV electricity – a cost that is by no means dictated solely by the choice of modules and the amount of sunlight. To understand this, we need to consider the capital and income components of a PV project.

Let us again imagine investing in a rooftop PV system. It is helpful to start by estimating expected *cash flows* over the life of the system, say 20 years, as in Figure 6.2. This is the key ingredient of what is known as *life-cycle analysis*.[1,2] Negative cash flows (expenditure) are shown red; positive ones (income) are shown blue. A major feature of PV systems is that the initial capital cost (*A*) produces by far the largest negative cash flow. This is followed by many years of positive cash flows representing the value of electricity generated (or savings due to electricity not purchased), and small negative ones to pay for routine system maintenance. Generally, it is also prudent to allow for additional capital expenditure to replace worn out or damaged BOS components such as charge regulators or inverters, or batteries in a stand-alone system (*B, C,* and *D*). And finally we may hope to obtain an end-of-life scrap value for the system (*E*).

We are now in a position to assess (perhaps with expert help!) the financial viability of the project. Of various measures, the easiest to understand are the simple *payback period*, the number of years it takes for the total costs

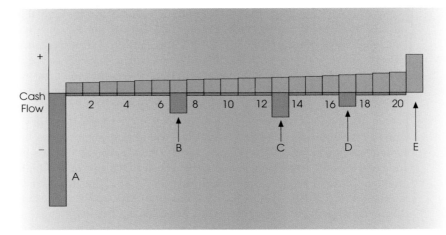

Figure 6.2 Positive and negative cash flows for a PV system.

to be paid for by the income derived from the system; and the *rate of return,* the percentage annual return on the initial investment. But it is hard to know how long the system will last, or to allow for additional capital injections that may be needed as time goes by (items *B, C* and *D* above).

An even more important limitation is that the simple payback period and rate of return take no account of the 'time value' of money – a major consideration for a long-term project. In a nutshell, a cash flow expected in the future should not be given the same monetary value today. For example, would you rather have € 100 today, or the expectation of € 150 in 10 years' time? Your answer will probably depend on predicting future interest rates (you could put the money in the bank); or the confidence you have about future payments; or you may prefer to purchase something for € 100 today. A proper life-cycle analysis takes this into account by referring all future cash flows to their equivalent value in today's money using a *discount rate.* This is the rate above general inflation at which money could be invested elsewhere, say between 1 and 5%. In this way the *present worth* of a complete long-term project can be estimated, and compared with alternatives, allowing a more realistic investment decision to be made. As you may imagine, a positive value of present worth is generally taken as a good indication of financial viability.

So far so good, providing we recognise that the decision, even when based on careful life-cycle analysis, contains uncertainties about technical performance, system and component lifetimes, interest rates, and the future

Figure 6.3 Investing in the future: PV for a school in South Africa (EPIA/IT Power).

price of electricity. And, as we have previously noted, it may also be based on environmental and social factors.

We have tried to summarise the ideas behind conventional life-cycle analysis, with its positive and negative cash flows. But what if the picture is clouded by a government decision to offer capital grants to offset the initial purchase price, or suddenly to change or terminate grants that are presently available? And what if the price paid for renewable electricity is bolstered by special tariffs that may be altered or removed by a change of government? Over the years there have been many such stop–go incidents in countries as wide apart as Australia, Spain and the USA. One of the biggest threats to rational decision-making and steady growth in the PV market is uncertainty about government policy; and one of the biggest benefits is consistent long-term support. We shall discuss support schemes in the next section.

You may be wondering why governments offer financial support to PV in the first place. There are two principal reasons. First, the products of a new

195

high-tech industry tend to be very expensive at the start, before cumulative production gathers pace. If governments wish to pursue urgent policy objectives such as the reduction of carbon emissions, they may decide to stimulate market development with financial incentives. Second, Figure 6.2 makes clear that PV, like other renewable energy technologies including wind and wave, has its major costs 'up front', with no fuel charges. This is quite different from conventional electricity generation based on fossil fuels. Projects with high initial costs that must be set against future income are commonplace for large corporations, but tend to be far more problematic for small businesses, organisations, and individuals who find it hard to raise the initial capital.

Government support, although generally welcome and necessary for PV, tends to distort the market and prevents it from behaving according to the assumptions of classical economics. Realistic life-cycle analysis becomes more problematic. In effect the global PV market becomes split into a number of sub-markets with different characteristics. As an extreme example, the decision of an organisation to install a large grid-connected system on its office building is likely to be influenced by very different financial criteria and incentives from that of a family in a developing country struggling to find initial funds for a solar home system. This is not to say that economic analysis is worthless, just that it should be approached and interpreted with caution. If you refer back to some of the photographs in earlier chapters, you will see plenty of examples of PV systems based on a wide range of investment criteria – political, economic, environmental, and social.

(a)   (b)

Figure 6.4 Diverse markets for rooftop PV systems: an elegant home in the developed world, and a 'mobile' home in Mongolia (EPIA/Shell Solar, EPIA/IT Power).

6.1.2 Financial incentives

We have already noted that PV, an exciting new technology with major environmental benefits, both justifies and deserves the support of governments wishing to accelerate market growth and counter the effects of global warming. Japan showed the way in 1994 with a 70 000 solar roofs program. Germany, after succeeding with its own 100 000 roofs program, went from strength to strength after 2004, thanks to improvements in its groundbreaking renewable energy legislation. Spanish government legislation led to an extraordinary burst of activity in 2008 when $2.7\,GW_p$ of PV capacity was installed in a single year (you may like to refer back to Section 4.5 on large PV power plants). The USA, held back during the years of the Bush administration, is now surging ahead. In spite of a certain amount of stop–go in all these programs, and difficulties due to the global economic recession that began in 2008, many other governments around the world have now joined the pioneers by offering substantial financial incentives to install PV systems.

Of the various ways in which governments have sought to provide financial incentives for the installation of grid-connected PV systems, two key ones are particularly relevant to our discussion here:

Figure 6.5 Rooftop arrays on the Reichstag building in Berlin exemplify the German government's support for PV (EPIA/Engotec).

- capital grants to offset the initial cost of the system.
- special tariffs for the electricity generated, which is either used on site or fed into the grid.

Referring back to Figure 6.2, the capital grant route is designed to reduce a project's initial negative cash flow, denoted by the letter A in the figure. Such grants, often covering 50% or more of the purchase price, are funded out of general taxation and are therefore paid for by all taxpayers. One disadvantage is that the money is paid up front, generally with no redress if the system is poorly maintained and fails to produce the expected amount of electricity. Another is that governments normally 'cap' the total amount of money available which can lead to an initial rush of grant applications that rapidly exhausts the fund – a perfect recipe for stop–go market development unless the scheme is constantly reviewed and reactivated.

The second approach, which offers attractive subsidies for electricity generated, increases the amount of income received over the life of the system (shown blue in Figure 6.2). It therefore encourages the purchase of high-quality systems that are carefully installed and maintained. Often taking the form of *feed-in tariffs (FITs)*, the subsidies are financed by requiring utilities to buy renewable electricity at well above normal market price. The cost is spread over all customers who must pay a small annual percentage increase in their electricity bills. From a government viewpoint FITs are generally 'revenue-neutral'. Their major advantage is the guaranteed income payments offered over timescales of 20 or 25 years, reducing uncertainty and increasing investor confidence.

PV systems that are designed to feed electricity into the grid must obviously incorporate appropriate metering. A one-way electricity meter to measure incoming power is no longer sufficient. One possibility is to replace it with a two-way meter that records the net flow to and from the grid, referred to as *net metering*. In effect the PV generator is paid the same rate per kWh for export and import, giving full value for all electricity produced. However FIT's and other schemes that pay differential rates for local generation (whether used on site or exported) require the PV output to be separately metered. The introduction of electronic smart metering in many countries will give greater flexibility in tariff design, allowing PV electricity to be priced according to the time of day it is generated and the requirements of the grid.

In recent years the FIT approach has proved increasingly popular, not least because of its remarkable success in Germany. A renewable energy law passed in 2000 introduced a FIT that proved extremely effective at stimulating a range of renewable energies. The PV tariffs were tweaked in 2004 to compensate for the termination of the German 100 000 roofs program, providing payback times of around 8–10 years. This resulted in a veritable boom in PV installations. Huge numbers of PV arrays were put on domestic and commercial buildings, farmers placed PV on barns and in fields, and many large PV power plants were commissioned. By 2005 total installed capacity in Germany exceeded $1\,GW_p$ and by 2008 it had reached $6\,GW_p$.

Of course, a generous FIT can become unsustainable if continued too long, so in many cases tariffs for new installations are lowered, or *degressed,* by a certain percentage each year to take account of PV's expected 'learning curve'. Providing they are well designed, such schemes avoid the need for caps on total capacity, and encourage suppliers to reduce costs and deliver more efficient systems. In 2008/09 there was much political debate in Germany about FIT tariff levels that had been standing at between 0.33 and 0.43 € /kWh according to the type and size of installation, and about possible caps. In short, the situation had become somewhat overheated and needed correction.

A Spanish FIT, first introduced in 1997, was upgraded in 2004, starting Spain on its exciting journey into the gigawatt era. A few years later the

Figure 6.6 This PV factory is in Malaga, Spain (EPIA/Isofoton).

Spanish government decided to introduce annual caps and slant the tariffs towards BIPV rather than large power plants, with ongoing reviews, to dampen a market that had surged beyond expectation.

More than 60 other countries have now entered the FIT arena and many are no doubt learning from the operational experiences of the pioneers. And in spite of the negative effects of the global economic recession that started in 2008, most commentators believe that PV and other renewable energy technologies will ride the storm relatively unscathed and continue to attract the support of governments increasingly focused on the dangers of global warming.

6.1.3   Rural electrification

So far we have been concentrating on economic aspects of grid-connected systems and the ways in which governments in developed nations encourage the development of PV markets. Passing reference has been made to stand-alone PV systems, noting that the chief competitor for supplying electricity in remote areas is generally the diesel generator. But all this relates to relatively wealthy nations including those that have driven PV's spectacular growth over the last decade.

There is another important dimension to the terrestrial PV story, and it concerns the provision of relatively small amounts of solar electricity to families and communities in the developing world, who have little prospect of buying and maintaining diesel generators, and no prospect of connection to a conventional electricity grid in the foreseeable future. This challenging yet worthwhile activity is referred to as *rural electrification*.[1,3]

A major aspect of rural electrification is the supply of *solar home systems (SHSs)* to individual families, and we shall concentrate on it here. Other applications include irrigation and water pumping (see Section 5.5.3), refrigeration of vaccines and medicines in remote hospitals, and the supply of PV systems to small businesses and institutions. It is sobering for those of us who live in developed countries to realise just how little electricity is needed to provide valuable services to people who otherwise have none. For example, the average electricity consumption of a household in Western Europe is around 10 kWh/day. The stand-alone system for a holiday home that we designed in Section 5.4.1 (see also Figure 5.15) assumed a consumption of 2.2 kWh/day, sufficient to run a good range of modern electrical appliances if used with care. But when we consider a SHS based on a single PV module, typically rated between 30 and 60 W_p, the figure is more likely to be 0.2 kWh/day – one-fiftieth of the electricity taken for granted

Figure 6.7 Pride of ownership: a family in China (EPIA/Shell Solar).

Figure 6.8 Selling solar in Kenya (EPIA/Free Energy Europe).

by most European families. This modest amount can power a few low-energy lights and a small TV, offering genuine improvements to rural living standards and a contact with the wider world.

Like other PV systems, SHSs have most of their costs up-front. A system comprising a small PV module, charge controller, 12 V battery, cabling, switches, and some low-energy lights may retail for a few hundred US dollars or euros. This may not seem much in America, Australia, or Europe, but to many families engaged in subsistence farming in the developing world it looks like an unattainable fortune. The SHS market – or perhaps we should say 'markets', because conditions vary widely from one country to another – therefore needs its own financing arrangements.

Efficient and convenient lighting is arguably the most important service offered by a SHS. Families without electricity often spend a substantial proportion of their disposable monthly income on kerosene lamps, candles, or dry cell batteries, so this money is in principle available to pay for a PV system.

Figure 6.9 PV modules and low-energy lights: a shop in Tibet (EPIA/IT Power).

Typical financing schemes for SHS's include:

- a short-term loan to cover all or most of the initial cost, paid back with interest over a period between 1 and 3 years.
- a leasing arrangement whereby a SHS is installed and maintained by an organisation or company in exchange for monthly *fee-for-service* payments.

A wide variety of official, commercial, and aid organisations are involved in the financing of SHS programmes around the world. In addition to national governments, local banks and leasing companies, the United Nations and the World Bank are actively involved and so are various non-government organisations (NGO's) and aid agencies.

Among the many countries with impressive rural electrification and SHS programs we might mention:

- *In Asia:* China, India, Sri Lanka, Bangladesh, Thailand and Nepal.
- *In the Americas:* Mexico, Brazil, Argentina, Bolivia and Peru.
- *In Africa:* Morocco, South Africa, Kenya and Uganda.

Figure 6.10 Enthusiasm for PV (EPIA/NAPS).

We end this section with a few comments on cultural and social issues surrounding the introduction of high-tech products into developing countries. In many cases a small stand-alone PV system represents the only contact of a rural family with 21st-century technology. Proper system maintenance can be a problem and education is a very important part of the package. Although PV modules are normally very reliable, the lead–acid batteries used in SHSs need regular topping-up and occasional replacement, modules must be kept reasonably free of dust and bird droppings, and electrical connections must remain tight and corrosion free. Such tasks are far removed from the experience of many rural communities. When SHSs are financed as part of a community electrification project there may be problems of management and control. A great deal has been learned over the past 30 years about the cultural pitfalls of rural electrification, where failures tend to occur for reasons other than the purely technical.[3] But any such difficulties should not detract from the great social benefits of rural electrification, which is surely one of PV's most admirable achievements.

Figure 6.11 Education, a very important part of the package (EPIA/Shell Solar).

6.2 Environmental aspects

6.2.1 Raw materials and land

The main environmental credentials of PV are established beyond doubt: its important contribution to reducing carbon emissions; cleanliness and silence in operation; lack of spent fuel or waste; and general public acceptability in terms of visual impact. We have already referred to such advantages at various points in this book. But there are further environmental considerations as PV accelerates into multi-gigawatt annual production – can Planet Earth provide the necessary quantities of raw materials, and is there enough land available for hundreds of millions of PV modules?

We start with the issue of raw materials. One point is clear: in so far as PV's future is based on silicon solar cells, there is no problem. Silicon is one of the commonest elements in the Earth's crust and, almost literally, as plentiful as sand on the beach. There is no future scenario in which it could become exhausted, and fortunately it is also essentially nontoxic. This is not to say that other materials involved in the manufacture of silicon PV modules are inexhaustible or problem-free, but silicon itself seems unassailable.

The situation is not so clearcut with the major new types of solar cell discussed in Chapter 2 – principally copper indium–gallium diselenide (CIGS), and cadmium telluride (CdTe). However a report published in 2004 by the highly respected National Renewable Energy Laboratory (NREL)[4], a facility of the US Department of Energy (DOE), was generally reassuring. The scenario considered was a rise in annual PV sales in the USA to $20\,GW_p$ by 2050, and the report estimated the requirements for 'specialty' materials needed to make the solar cells, additional materials to imbed them in PV modules, and 'commodity' materials for such balance-of-system (BOS) items as roof mountings and support structures. It then compared the amounts of the various materials with current global production levels and estimated the percentage annual growth required until 2050. It concluded that the above scenario would not create problems with materials availability, although the situation could change if growth proceeded much more quickly, or if world production were to reach $100\,GW_p/year$.

Although the picture has changed somewhat since 2004, and will no doubt change again in the coming years, some of the report's main findings remain helpful. They are summarised in Figure 6.13 using colour-coding to denote the degree of 'supply constraint' in the various materials needed

Figure 6.12 Effectively inexhaustible: silicon for solar cells (EPIA/Solar World).

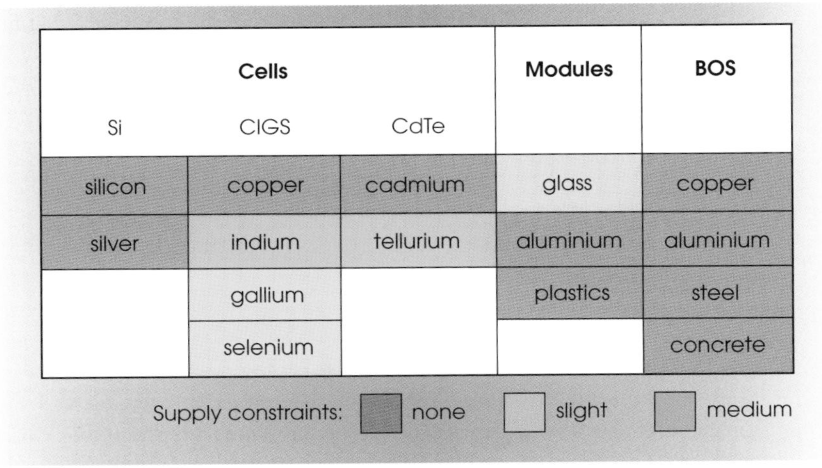

Cells			Modules	BOS
Si	CIGS	CdTe		
silicon	copper	cadmium	glass	copper
silver	indium	tellurium	aluminium	aluminium
	gallium		plastics	steel
	selenium			concrete

Supply constraints: ▇ none ☐ slight ▢ medium

Figure 6.13 Supply constraints on major PV materials.

207

to satisfy the projected $20\,GW_p$/year demand. As we might expect, silicon solar cells are given the all-clear, with no concern about either silicon itself or the silver used to screen-print cell interconnections. CIGS cells might be slightly constrained by shortages of gallium and selenium, and more so by shortages of indium; CdTe cells by a shortage of tellurium. Moving on to PV modules and BOS components, the only slight concern is over the large amounts of glass required – not because the raw materials (more sand!) would run out, but because global production capacity would have to rise substantially to meet PV's demands. In conclusion, the only real concerns are for the 'specialty' materials indium and tellurium, and to a lesser extent gallium and selenium.

The report wisely notes a number of uncertainties, and suggests strategies and developments that would mitigate shortages of key materials. It is, of course, unlikely that CIGS and CdTe cells will continue to be made exactly on today's pattern – for example, the active layers may become much thinner. It is probable that other types of cell currently in the research phase, or entirely new ones not yet discovered, will be in volume production by 2050; and in any case the calculations were based on the unlikely assumption that all the required $20\,GW_p$ would be met by one of the existing thin-film PV technologies – not including any contribution from silicon!

Yet there are some significant, and ongoing, concerns, especially about indium and tellurium. In the past few years there have been scare stories in the press, notably about indium, and its price on world markets has fluctuated wildly. To a large extent the problem has arisen because of its use in liquid crystal display (LCD) monitors and TV screens, an application consuming up to 50% of world production that could not have been foreseen 20 years ago. This surely highlights the difficulties of predicting future availability of important materials that are constantly finding new applications, many of them far removed from PV, and fading away from old ones. Who knows what the next 20 years will bring, let alone the next 40?

A general point worth making is that the solar cell materials presently seen as potential bottlenecks are byproducts of major mining operations. Indium is a byproduct of zinc extraction, tellurium (and selenium) of copper extraction. In the normal course of things, the amounts of these byproducts fluctuate in sympathy with production levels from the main mining operations. Interestingly, indium is not an especially rare element in the Earth's crust; it is actually about three times more plentiful than silver, but only extracted at one-sixtieth the rate, emphasising the dependence of indium volumes on the scale of zinc mining. Experts make the point that increasing scarcity of a byproduct, inevitably reflected in the market price, tends to encourage more careful processing of the parent ore. It also stimulates recycling, which has recently been satisfying up to half of the demand for indium, and the search for alternative materials.

Gallium is another material considered in the NREL report because of its use in CIGS cells. In recent years gallium has branched out in the form of gallium arsenide (GaAs) cells for space vehicles and, potentially more important from the availability viewpoint, into high-concentration terrestrial PV modules (see Sections 2.4.3.1 and 3.4). GaAs is also in line for application in other fields including a future generation of very high-speed computer chips, replacing silicon. So gallium may be moving slightly higher on the list of supply constraints.

How does the overall situation compare with the scenario painted by the NREL report in 2004? The huge recent rise in global PV production, and the promise of further rapid growth to 2020 and beyond, certainly bring into focus the report's caveat about global thin-film production exceeding $100\,GW_p$ per year. Yet thin-film still accounts for less than 10% of global production, and much of the current surge is coming from new factories in China that manufacture wafer-based silicon PV modules. So it will be many years before thin-film manufacturing overtakes crystalline silicon, and even then it will certainly not be based on a single technology. Amid the rather confusing debates about specialised PV materials there are grounds for cautious optimism, especially given the ability of the PV community to innovate and adapt. And of course there are plenty of silicon enthusiasts who can afford to watch from the sidelines, ignoring all talk about scarcity of raw materials!

We now turn to the question of land use. This has already been mentioned in Section 1.5, where we suggested that an area of land $140 \times 140\,km$, or $20\,000\,km^2$, roughly three times the size of London or Paris, would be sufficient to accommodate 1000 GWp of PV modules. It seems that by 2020, or soon after, we may be approaching this huge total, some 50 times greater than global installed capacity in 2009, assuming PV continues its present remarkable progress. But where would the land actually come from, and would we resent it?

If $20\,000\,km^2$ sounds like a large parcel of land, consider some even larger ones: the Sahara Desert is about 850 times bigger; the Australian Outback about 200 times; and the state of Arizona about 15 times. In the USA, cities and towns cover some $700\,000\,km^2$ and in many countries wide tracts of land are set aside for military uses, airports, highways, fuel pipelines, and so on. In short, if the world's PV is sensibly spread around among the

Figure 6.14 No need for extra land: a rooftop PV array at Munich Airport (EPIA/BP Solar).

world's nations, the landscapes seen by the vast majority of people will be virtually unchanged from those they enjoy today.

Of course this is far from the whole story, because PV can be installed on buildings. There are vast numbers of existing homes, offices, public buildings, factories, warehouses, airports, parking lots and railway stations with suitable roofs and façades, and we may be sure the that tomorrow's architects will be even more aware of the possibilities. BIPV will undoubtedly provide a major part of PV's future space requirements, leaving deserts and other unproductive land to supply most of the balance. Sunshine is everywhere, high and low, city and country, and at fairly predictable levels. There is absolutely no need for PV to dominate with unsightly and unwelcome 'blots on the landscape'.

6.2.2 Life-cycle analysis

In the previous section we considered PV's requirements for raw materials and land – two environmental issues that surface before PV production even begins. Further important environmental questions arise during a PV system's lifetime, which starts with extraction and purification of raw materials; proceeds through manufacture, installation, and many years of operation; and ends with recycling or disposal of waste products. The whole sequence is referred to as a *life cycle*, and it is important to appreciate its environmental consequences. Note that this form of life-cycle analysis (LCA) is not the same as the classical economic version introduced in the previous section, which deals with cash flows and financial decisions. We are now moving on to something much broader, with important implications for global energy policy and society as a whole.

In this brief introduction we will consider LCA under two main headings:

- *Environmental and societal costs.* What costs, in addition to classic economic costs, are incurred or avoided?

- *Energy balance.* How does the amount of electrical energy generated over a system's lifetime compare with the energy expended in making, installing, and using it?

We start with environmental and societal costs.[1] It is clear that all methods of energy production – whether based on oil, gas, coal, nuclear, or renewable sources – have impacts on the environment and society at large that are ignored by the traditional notion of 'cost'. A narrow economic view of industrial processes assesses everything in terms of money, while ignoring other factors that common sense tells us should be taken into account in any sensible appraisal of value. For example, the 'cost' of generating electricity in nuclear power plants has traditionally been computed without taking any account of accident or health risks; in the case of coal-fired plants, without acknowledging their unwelcome contribution to global warming; and with wind power, without placing any value on landscape.

There are two main reasons for this apparent short-sightedness. First, aspects such as health, safety, environmental protection, and the beauty of a landscape cannot easily be quantified and assessed within a traditional accounting framework. We all know they are precious, and in many cases at least as important to us as money, but appropriate tools and methodologies for including them are only now being developed and accepted. It is surely vital to do this, because so many of our current problems are bound up with the tendency of conventional accounting 'to know the price of everything and the value of nothing'.

The second reason relates to the important notion of the *external costs* of energy generation. These costs, most of which are environmental or societal in nature, have generally been treated as outside the energy economy and to be borne by society as a whole, either in monetary terms by taxation, or in environmental terms by a reduction in the quality of life.[1] They contrast with the *internal costs* of running a business – for buildings and machinery, fuel, staff wages and so on – that are paid directly by a company and affect its profits. If Planet Earth is treated as an infinite 'source' of raw materials and an infinite 'sink' for all pollution and waste products, it is rather easy to ignore external costs. For example it seems doubtful whether the 19th-century pioneers of steam locomotion ever worried much about burning huge quantities of coal; or the 20th-century designers of supersonic civil airliners about fuel efficiency and supersonic bangs. One of the remarkable changes currently taking place is a growing world view that external costs should be worked into the equation – not just the local or national equation, but increasingly the global one. In other words external costs should be *internalised* and laid at the door of the responsible industry or company. In modern phraseology, 'the polluter should pay'.

Many of the external and internal costs associated with industrial production are illustrated by Figure 6.15. The external ones, representing charges or burdens on society as a whole, are split into environmental and societal categories, although there is quite a lot of overlap between them. You can probably think of some extra ones. Internal costs, borne directly by the organisation or company itself, cover a very wide range of goods and serv-

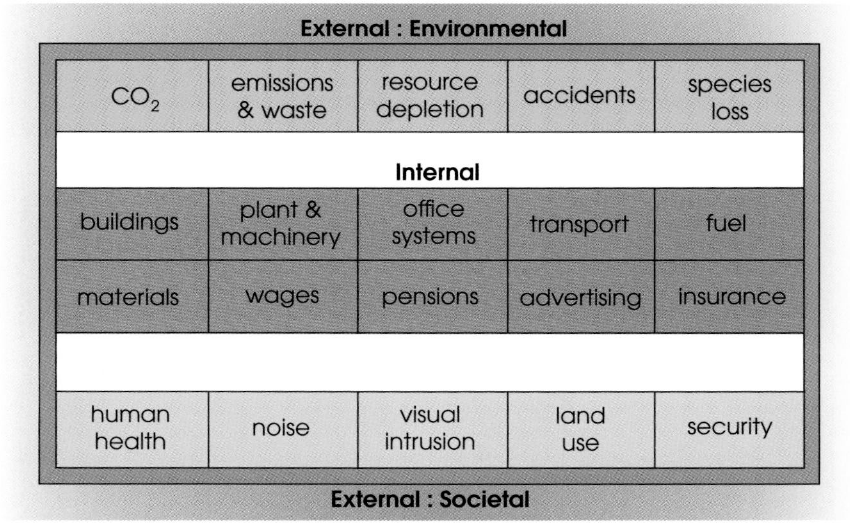

External : Environmental				
CO_2	emissions & waste	resource depletion	accidents	species loss
Internal				
buildings	plant & machinery	office systems	transport	fuel
materials	wages	pensions	advertising	insurance
human health	noise	visual intrusion	land use	security
External : Societal				

Figure 6.15 External and internal costs.

ices, from buildings to staff wages. The distinction between internal and external costs is somewhat clouded by the fact that many items bought in by a company, for example fuel and materials, have themselves involved substantial 'external' costs during production and transport. In the case of electricity generation a proper analysis of the environmental burdens should take proper account of all contributing processes and services 'from cradle to grave', whether conducted on- or off-site. Needless to say this is a challenging task.

One of the special difficulties facing renewable electricity generation, including PV, is that so many of its advantages stem from the *avoidance* of external costs and are therefore hidden by conventional accounting methods. Renewables tend to produce very low carbon dioxide emissions, cause little pollution, make little noise, create few hazards to life or property, and have wide public support. PV can claim all these benefits. Yet when economists and politicians talk about PV, reduction or avoidance of external costs is seldom mentioned. Fortunately, energy experts and advisers to governments are taking increasing notice of environmental life cycle analysis in their decisions, and assessing the risks and benefits of competing technologies on a more even footing.[5] Certainly, the PV community must be involved in countering outdated thinking about the wider benefits of its technology.

We now move on to the much-discussed topic of *energy balance*. Clearly, it takes energy to produce energy. But how does the total amount of electrical energy generated by a PV module or system over its lifetime actually compare with the input energy used to manufacture, install, and use it? Closely related to the energy balance is the *energy payback time,* the number of years it takes for the input energy to be paid back by the system.[6] We naturally expect PV to have favourable energy balances and payback times, especially in view of its claims to be clean and green.

Two initial points are worth making. First, energy payback is not the same as economic payback. The latter is concerned with repaying a system's capital and maintenance costs (including cost of energy consumed) by a long-term flow of income, and is essentially a financial matter; energy payback is much more about the environment. Secondly, the environmental benefits of a short payback time depend on the present energy mix of the country, or countries, concerned. If the required input energy is largely derived from coal-burning power plants, it is more damaging than if it comes from, say, hydroelectricity.

Major energy inputs to a PV system occur during the following activities:

- extraction, refining, and purification of materials.
- manufacture of cells, modules, and BOS components.
- transport and installation.

Interestingly, some of the most significant energy inputs are for components such as aluminium frames and glass for modules, and concrete foundations for support structures in large PV plants. Although the energy required to refine pure silicon and make crystalline silicon solar cells is considerable, the continual trend towards thinner wafers using less semiconductor material is reducing this problem. The energy input for thin-film cells is generally very small.

The other side of the energy balance – the total electrical energy generated by a system over its lifetime – depends on a number of factors discussed in previous chapters:

- efficiency of PV modules and other system components.
- the amount of annual insolation.
- alignment of the PV array, and shading (if any).
- the life of the system.

The energy balance is most favourable for systems that are efficiently produced in state-of-the-art factories, and installed at optimal sites in sunshine countries. Things get even better if systems last for longer than their projected or guaranteed lifetimes – but of course this is hard to predict.

Some major life-cycle studies carried out in the early years of the new millennium painted a rather gloomy picture of PV's environmental and health impacts, due largely to the fossil-fuel energy used during cell and module manufacture. However, a more up-to-date report[5] that takes proper account of external costs and recent advances in PV engineering comes to far more optimistic conclusions.

The report considers the many factors, including rising solar cell efficiencies, use of thinner semiconductor layers, larger more energy-efficient factories and processes, and improvements in BOS components that are driving down PV's environmental impacts year by year. Energy payback times for roof-mounted systems are especially favourable because of their modest BOS requirements, including light mounting structures. If based on crystalline silicon modules and installed in southern Europe or sunshine states of the USA with typical annual insolation values of around $1700 \, kWh/m^2$, payback times are currently about 2 years – surely an excellent result for systems expected to last for 25 years or more. The figure is nearer 4

Figure 6.16 Helping reduce environmental impacts: a modern solar cell factory (EPIA/Q-cells).

years for similar systems installed in the less sunny climates of northern Germany, the Netherlands, or the UK. The situation is even better for systems based on the new generation of thin-film modules, which use extremely small quantities of active semiconductor materials. In favourable locations energy payback may soon take no longer than a single orbit of Planet Earth around the Sun – another pointer to PV's exciting future.

References

1. T. Markvart (ed.). *Solar Electricity* (2nd ed), John Wiley & Sons, Ltd: Chichester (2000).
2. R.A. Whisnant et al. *Economic Analysis and Environmental Aspects of Photovoltaic Systems*, in A. Luque and S. Hegedus (eds.), *Handbook of Photovoltaic Science and Engineering*, John Wiley & Sons, Ltd: Chichester (2003).

3. J.M. Huacuz and L. Gunaratne. *Photovoltaics and Development*, in reference 2.
4. National Renewable Energy Laboratory (www.nrel.gov). *Will we have enough materials for energy-significant PV production?* Report DOE/GO-102004-1834 (2004).
5. V.M. Fthenakis and H.C. Kim. Quantifying the Life Cycle Environmental Profile of Photovoltaics and Comparisons with Other Electricity-Generating Technologies, *IEEE 4th World Conference on Photovoltaic Conversion*, Hawaii (2006).
6. E.A. Alsema. Energy Pay-back Time and CO_2 Emissions of PV Systems, *Progress in Photovoltaics* **8**, 17–25 (2000).

Index

Electricity from Sunlight   By Paul A. Lynn
© 2010 John Wiley & Sons, Ltd